The Ethics of Beekeeping

by John M Whitaker

Northern Bee Books

The Ethics of Beekeeping
by John M Whitaker

ISBN 978-1-912271-24-5

Published by Northern Bee Books, 2018
Scout Bottom Farm
Mytholmroyd
Hebden Bridge HX7 5JS (UK)

Design and artwork by JEH

Printed by Lightning Source UK

Contents

4

1 – Ethics

A few years ago, I spoke to a group of ladies about beekeeping. They had listened with interest and at the end of the talk had asked several interesting and perceptive questions. It was mentioned that one lady who belonged to their group had chosen not to attend the meeting as she didn't believe in the ethics of beekeeping. I was rather surprised at this. We beekeepers tend to think we are the good guys, that without beekeepers, honey bees could become an endangered species. Our bees are one of the most important pollinators of flowers, and they have enormous significance in the production of many foodstuffs.

This exchange lingered in my mind and the more I considered it, I realised how complicated and many faceted were the ethics of beekeeping. Beekeeping is beset by choices and choice by its very nature could involve, to a greater or lesser extent, ethics. This small book is trying to clarify where, in beekeeping, there are ethical choices to be made. It is not about trying to persuade the reader of the rightness or wrongness of one choice as opposed to another. Nor is it an exposition of what I personally think is right or wrong in beekeeping, though it is difficult, indeed impossible, to completely hide my own beliefs. The choices are complicated as they impinge not just on the beekeeper, but also on the welfare of the bees, other beekeepers and society in general. Due to the very nature of ethical choices, there will be contradictions and counter arguments. This is not a book to tell you how to keep bees ethically. How you chose to keep your bees is your decision. This book simply rehearses for the reader the ethical arguments that may arise when keeping bees. Whatever choices you make, there will inevitably be compromises. If you are looking for absolute values in life do not keep bees.

One of problems I have had is deciding who this book is aimed at. Is the reader likely to be a new beekeeper, an experienced beekeeper or simply someone who is interested in bees and beekeeping? Or possibly nobody? Eventually I decided to be inclusive - the book is for anyone who wishes to consider the ethical side of beekeeping. This does mean that, for the non-beekeeper, I have needed to include more detail than the experi-enced beekeeper might require.

My understanding is that ethics is the study of whether an action or intention is right and wrong. When an action is termed ethical this means that it is right according to an accepted code of ethics. If an action is termed unethical it is wrong. But this raises the question of how one arrives at a judgement of what is right and wrong. Is a code of ethics something that is absolute, or does it vary with time and place?

Human beings often claim to be more highly evolved than other creatures. This is blatantly not the case. After all, all creatures, including humans, have evolved over the same period, that is the 3500 million years since the first organisms that are based on DNA appeared on earth. We do not even have the most complex DNA and many of the characteristics of our phenotype, are quite obviously inferior to those of some other animals. Many animals are stronger, can run faster and have better eyesight than human beings. But we do possess a remarkable brain that gives us a prodigious memory, and the ability to reason, plan and envisage a future. And significantly, as we grow into adulthood, we develop an awareness of the effects of our actions on other creatures. We are not born with this. Toddlers can be remarkably unaware of the pain they can cause to fellow toddlers and even their mothers. They may bite adults close to them, strike other children and torment pets. But usually by the time they are ready for school parents will have taught them that violence is not acceptable in the family circle and with their friends and as they learn to speak they also learn basic manners. This is the start of learning ethics, not just that their behaviour can affect other people but that behaviour can be categorised as either right or wrong. The development of personal ethics is a process that continues most of our lives and certainly into early adulthood. In Jane Austen's 'Emma', Emma, clever, vivacious and witty and in many ways an attractive young woman allows herself to mock the ridiculousness of an ageing spinster to her face. She hadn't intended to be deliberately cruel but that was the effect. Afterwards, her long time friend and mentor, Mr Wrightly took her to task telling her 'That was badly done Emma' and in that moment she realised the shortcomings in the way she related to her companions and friends.

Although ethics is largely a learnt behaviour, there is also evidence that chimpanzees and other animals have evolved empathy, rule governed behaviour and some researchers suggest that altruistic behaviour has evolved in some animals. Neuroscientists are suggesting that there are areas within the human brain responsible for ethical thinking.

Most religions have at their core a set of laws which form a basis for ethics for individuals within society. Despite the disparity between the dogmas of the major religions there is much similarity between the ethics that different religions promote.

The laws of all major religions preach a respect of life and property, especially amongst those within their own specific social grouping. The old testament gave us the ten commandments and Jesus simplified these to two commandments. As religions became more widespread these principles were extended to generally include all members of a nation and then groups of nations. But it is evident that these principles are not universally adopted even amongst those that profess the major religions of Christianity, Islam, Judaism or Hinduism, as can be seen by what is currently happening in many parts of the world. Secular movements do not always have a better record. The democratic nations have adopted a code of law and in recent times a bill of human rights, extending the rights of individuals within a society far beyond the basic respect of life and property. Communist and totalitarian states likewise have a code of behaviour set out for their citizens. Whereas the democracies design their codes on the basis of the freedom and equality of the individual members of their society, even where this can compromise the health of the society as a whole, totalitarian states base their code of behaviour on what is good for the society as a whole, even where this will compromise the rights of the individual. But both systems find it difficult to extend their code of behaviour outside of national borders. It is this difficulty that has lead to the creation of the United Nations and the European Union.

When rights are granted to individuals, it follows that a code of ethics is created, as there is now a right and wrong way to behave way with respect to our fellow citizens. So our ethics, our sense of right and wrong, are fashioned by philosophy, by social attitudes, by religion, and by the secular laws of the society in which we live.

Over the centuries philosophers have wrestled with the problem of how to base ethics upon a rational and logical foundation. Aristotle, two and a half millennia ago, tried to base ethics upon the essence or purpose of human life and this, many centuries later, was developed by Thomas Aquinas into the 'natural law' approach to moral questions. Aquinas argued that everything, whether material or process, having being created by God, was given by God an ultimate purpose. Human beings, having been given reason, can strive to understand God's ultimate purpose and if in their actions they pursue this ultimate purpose then this action is good, and if their actions frustrate or run counter to this ultimate purpose then this is bad. Immediately it is possible to see that this raises difficulties when we consider the ethics of beekeeping. What is the ultimate purpose of bees? Is it to produce honey, or is it to pollinate flowers or is it to simply survive? Each of these ultimate purposes could lead to different conclusions as to how we keep bees.

The thinking of Thomas Aquinas still significantly influences the teaching of the Roman Catholic church, particularly in arena of sexual behaviour. For instance if the ultimate purpose of sex is to fertilise an egg and so produce another human being, then it follows that sexual intercourse between members of the same sex and sexual intercourse between those who are outside childbearing age are both wrong because neither action can lead to conception. These conclusions are widely disputed in much of current thinking in western society. The concept of 'natural law' generally, but not necessarily, presupposes the existence of a omnipotent God who created everything. Also, once an ultimate purpose is established it produces a rule of behaviour that can be applied to all people at all times. According to the principle of Natural law, the fact that an act, which is deemed right, may have unintended consequences which is harmful does not affect the rightness of the act. So that an act of sexual intercourse between a man and women, leading to the birth of a child which cannot be provided for, does not invalidate the rightness of the act.

Utilitarianism.

Jeremy Bentham put forward the idea of utilitarianism, whereby an action should be judged by the net results it achieves. An action is deemed to be good if it results in more benefit than harm to the people affected by the action. This may seem like common sense but in fact it is virtually impossible to measure benefit and harm to individuals and so there is no easy way to implement the concept of utilitarianism in practice. The current stress on human rights means that the harm to one individual is usually judged to be of more significance than a smaller benefit to a large number of persons. The ideas of utilitarianism seem to be very much in accord with Darwin's theory of evolution, but, while Darwinian ideas are accepted in science to explain the evolution of all living species, they are not accepted by political thinkers in their consideration of how society will progress. This too is not easy to apply to beekeeping as the net benefit must take into account the effect on the bees, beekeepers and society in general.

Legal framework and ethics.

Parallel to the philosophical and religious basis of ethical codes, during the last two and a half millennia, nation states have developed legal codes. If a ruler, or ruling body, is to properly control its regime there must be an accepted code of behaviour, and when this code of behaviour is enforced by punishment it becomes the law. It was the Roman city state, centuries before it became an empire that dominated Europe, that recognised that if law is to function then it must be written down and codified so that those that administer the law can do so consistently and fairly. It was the Roman system that differentiated between the criminal and civil law. The fundamentals of Roman Law still form the foundation of the legal systems in much of Europe, but in northern Europe and in particular England, the present day legal system is based on common law and the precedents generated by generations of judgements.

The law and ethical codes are not the same thing but they do have much in common. In some instances, the law and ethics overlap and what is perceived as unethical is also illegal. In other situations, they do not overlap. In some cases, what is perceived as unethical is still legal, and in others, what is illegal is perceived as ethical. It could be considered that the law sets minimum standards of behaviour while ethics sets maximum standards.

Laws must be consistent, and it should not be possible to have two requirements that are contradictory. A law must be universal and apply to everyone who is subject to it. A law needs to be codified, and published so that it is known to all its subjects and so can be consistently administered. Though it wasn't always the case, it is now thought that a law must be generally accepted so that it is obeyed by the majority without dissent. A law needs to be enforced and society must be prepared to punish those that choose to not obey the law.

Ethics involves learning what is right and wrong, and then doing the right thing. Most ethical decisions have multiple consequences, mixed outcomes, uncertain consequences and personal implications.

In the first instance the development of a legal code reflected the accepted ethics of the day. However, to some extent, the introduction of new laws can alter accepted ethics. For instance, in 1983 a law was passed in the UK making the wearing of front seat belts compulsory and in 1989 it became compulsory to wear seat belts in the rear seats of a car. When these laws were introduced they were not universally accepted. Thirty years later most people believe that wearing seat belts is the right thing to do, and would argue that it is not just a matter of your own safety while travelling but also in the interests of your fellow passengers and other road users.

Animal Rights.

Ethics as it relates to the treatment of animals has, in the main, only became well defined in the last century. Before that, attitudes towards animals were mainly defined by the predominant religions. The sacred texts of all three of the major world religions do address the rights of animals, but do not give the subject great weight.

Islam teaches that all living things were created by Allah and that Allah loves all animals. Though animals exist for the benefit of human beings, animals must be treated with kindness and compassion. Muslims are taught to avoid cruelty to animals, in particular overworking animals, neglecting animals, hunting animals for sport, animal fighting for sport and factory farming.

Whether all the above strictures are derived directly from the Qur'an is not clear but the Qur'an does contain a number of stories that illustrate the concern of the Prophet for the welfare of animals. For instance when the Prophet was asked if Allah rewarded acts of charity to animals, he replied: "Yes, there is a reward for acts of charity to every beast alive." Elsewhere the Prophet said "Whoever kills a sparrow or anything bigger than that without a just cause, Allah will hold him accountable on the Day of Judgment." On the other hand the Qur'an explicitly states that animals may be exploited for human benefit provided this does not entail unnecessary suffering. Qur'an 40;79-80 'It is God who provided for you all manner of livestock, that you may ride on some of them and from some you may derive food. And other uses in them for you to satisfy your heart's desires. It is on them, as on ships, that you make your journeys.'

In the past Christianity and the Jewish tradition from which it developed, have been largely indifferent to the suffering of animals. Christian theologians believed that human beings were greatly superior to animals. The bible appears to impose few moral obligations for humans as regards animals, on the basis that humans have souls and animals don't and humans have reason and animals don't.

Taking a quote from Genesis 1:26 'Then God said, "Let us make mankind in our image, in our likeness, so that they may rule over the fish in the sea and the birds in the sky, over the livestock and all the wild animals and over all the creatures that move along the ground."

However, despite the few strictures in the bible to support animal rights, in recent years Christianity has supported animal rights, and argued against animal cruelty. There are sects within Christianity that are strict vegetarians. Modern Christian thinking is largely sympathetic to animals and less willing to accept that there is an unbridgeable gap between animals and human beings. Although most theologians don't accept that animals have rights, they do acknowledge that some animals display sufficient consciousness and self-awareness to deserve moral consideration. The growth of the environmental movement has also radically changed Christian ideas about the role human beings play in relation to nature. Few Christians nowadays think that nature exists to serve humanity, and there is a general acceptance that human dominion over nature should be seen as stewardship and partnership rather than domination and exploitation and consequently there has been a significant softening of Christian attitudes to animals.

Of the three main world religions, it is Hinduism that histori-cally was the most concerned about animals and regards human beings and animals as equal creatures upon earth. Many Hindus are vegetarians. In 'The moral basis of vegetarianism ' Mahatma Gandhi states 'The greatness of a nation and its moral progress can be judged by the way its animals are treated.'

The secular governments of the world, responding to the concerns of citizens, slowly began to introduce legislation to protect animals from unnecessary cruelty. To start with the acts passed by parlia-ment concerned specific groups of animals, but over the years, subsequent legislation arrived to supersede what had been passed before and expanded the range of animals that were protected. In many ways the UK was at the forefront of this development. One of the first of acts passed by the UK parliament was the Cruel Treatment of Cattle Act 1822. This was followed by The Cruelty to Animals Act 1849, The Cruelty to Animals Act 1876, The Protec-tion of Animals Act 1911 and finally The Animal Welfare act 2006, which is currently in force in the UK.

When the EU Lisbon Treaty came into force in 2009 it introduced the recognition that animals are sentient beings. Article 13 of Title II states that:

"In formulating and implementing the Union's agriculture, fisheries, transport, internal market, research and technological development and space policies, the Union and the Member States shall, since animals are sentient beings, pay full regard to the welfare requirements of animals, while respecting the legislative or administrative provisions and customs of the Member States relating in particular to religious rites, cultural traditions and regional heritage."

The introduction of the concept of sentience in animals, and above all the acceptance that animals feel pain, has resulted in the utilitarian concept of ethics being moulded to become much more about the minimisation of pain rather than the maximisation of pleasure.

2 – The Honey Bee Nest

In this chapter the issues are discussed that arise when honey bees are placed in a situation that differs from that which they experience in the wild.

When beekeepers choose to keep honey bees, they are taking on a responsibility, a duty of care towards them in the same way as when we choose to look after farm animals or domestic pets. Whereas most animals that we might look after have, to some extent, been domesticated and have adapted over many generations to human management, honey bees are still, in most respects, feral creatures. Beekeepers are taking them out of their natural environment and putting them, for the convenience of the beekeeper, in an environment of the beekeepers' choice, an environment which, if the bees were left to their own devices, they might not choose. When they take bees from their natural environment, beekeepers must be prepared to take on responsibility for the welfare of their honey bees.

In the previous sentence I have casually used the word 'their', implying that the beekeeper owns the bees. But ownership of bees is complicated. The ownership only applies when the colony of honey bees agrees to continue to live in the hive that the beekeeper provides for it. If the bees abscond or swarm on to someone else's property that ownership no longer exists, either in law or practice.

In the wild, a colony of honey bees would generally nest in a natu-

Exposing a feral colony that has been discovered in a bricked up window.

ral cavity. Typically this would be within a hollow trunk of a tree, but equally it could be in a cavity in a rock face. With the arrival of human beings, almost 40 million years after bees evolved into their current form, alternative cavities became available in the stone walls of buildings or the eaves of houses.

When they swarm, honey bees go to elaborate lengths to select and agree amongst themselves upon the cavity that they will occupy. It should be neither too big nor too small. On average they select a cavity about forty litres in volume. Preferably it should have a single small entrance towards the base of the cavity facing south and at a height that most predators would be unable to reach. To some extent honey bees are able to modify these cavities using propolis, a natural resin collected from plants, blocking up unnecessary entrances and possibly reducing the size of the entrance.

Within this cavity the honey bees build a structure of honey combs. These are built from wax that is secreted from the eight wax glands found on the ventral side of the abdomen. The combs adhere to the ceiling of the cavity, the anchor being reinforce with propolis. Each comb is a lamina structure, hanging vertically from its anchor on the ceiling. The number of combs will depend upon the shape of the space in the cavity. The combs are separated from each other sufficiently so that there is space for the bees to move on the surfaces of adjacent combs, working back to back. This means that the distance between the combs, centre to centre, is normally between 35mm and 38mm. On both sides of the wax comb the bees build a tessellation of hexagonal cells. At first glance it appears that the axis of each cell is horizontal but closer examination shows that each cell angles upwards away from the central septum at 13o. The majority of the cells are between 5.2mm and 5.4mm in diameter. A smaller number, often at the periphery of the comb, are between 6.2 and 6.4mm in diameter. The walls of the cells are uniformly 0.07mm in thickness. The smaller and more numerous cells are used to raise worker brood, the larger ones to raise drone brood. Both types of cells may also be used to store pollen or honey.

The combs hang from the ceiling of the cavity. To give additional strength the sides of the upper part of the comb may also adhere to the sides of the cavity. The lower part of the comb hangs free, with the bottom edge forming a catenary curve (the curve formed by a hanging chain). Queen cells are produced, usually in May and June, when the colony needs to raise a new queen. Queen cells, uniquely, hang vertically and are often built on the sides or the lower edge of the comb, though they can also be built on the surface of the comb.

The honey bee nest has a typical structure. Though it is usually viewed by examining single combs which give a cross sectional view of the nest, it really needs to be imagined in three dimensions. At the centre and towards the lower part is the area used to raise brood. When looking at a cross section by examining a single comb it can be seen that the brood nest was made up of concentric areas of eggs, then larvae, then sealed brood. As the comb is built downwards the centre of the brood nest moves downward. It is observed that the queen prefers to lay in the newly formed wax comb. Drone brood is found at the extremes of the brood nest, but usually just for a limited period from April until August. Pollen is stored around the sphere of brood nest, and then above and to the sides is where the honey stores are usually placed. Whenever one is describing any aspect of honey bee behaviour it is necessary to qualify what is said using the words 'usually' or 'normally'.

The structure of the nest results in air being confined between the combs, with little natural flow. So, besides providing a place to raise the brood and store honey and pollen, the combs produce an environment in which the temperature and humidity can be controlled by the bees. The air between the combs can either be moved and circulated by fanning, that is by the bees flapping their wings, or it can be trapped using the bodies of the bees to inhibit the movement of air. During the winter months or cool nights in the summer, honey bees are said to cluster, inhibiting air flow and so retaining the heat within the colony. Though the term 'cluster' is in common use in this context I find this term rather misleading as it doesn't fully describe the essential contribution made by the wax combs.

Regardless of the external conditions, the bees must maintain the environment in the area where they are raising their brood at a temperature about 34oC to 36oC. When the brood temperature needs to be raised, the worker bees can generate additional heat by flexing their flying muscles without flapping their wings, convert-ing the energy in the honey they are eating into heat energy. When the brood area temperature needs to be reduced the workers can fan to move the hot, and often humid, air out of the hive. In more extreme high temperatures the bees will collect water, spreading it

A national brood box with moveable frames.

as a film on the combs and allow the water to evaporate removing yet more heat from the hive, just as we cool ourselves by perspiring. This ability to maintain a constant temperature within the hive is known as thermoregulation and is achieved by a combination of their ability to make wax comb and various behaviours. As a result of this, honey bees are able to thrive in a large range of climatic conditions, from the tropics to temperate zones, provided nectar and pollen are available. The ability of a species to thrive over such a large proportion of the earth's land surface is rare.

The use of moveable frames within hives followed a patent submitted by the Rev L L Langstroth in 1852. It revolutionised beekeeping, bringing several advantages to those wishing to keep bees and harvest the honey. When the beekeeper uses moveable frames, the honey can be easily harvested without damaging the brood area or harming the bees. The development of the colony can be monitored, so that disease, should it occur, can be spotted at an early stage and remedial steps taken. Preparations for swarming can also be observed and then measures taken to control the process so that the bees are not lost. And finally, the ability to view closely the inner workings of a colony of honey bees opened up a fascinating world to thousands of ordinary people, as well as to researchers and scientists. Honey bees became the most studied and written about insect in history.

The use of wooden frames within the hive results in significant changes to the environment within the hive. The wax of the comb is attached to the frame at the sides, top and bottom. Surrounding the frames, at the sides, top and bottom is a gap which is known

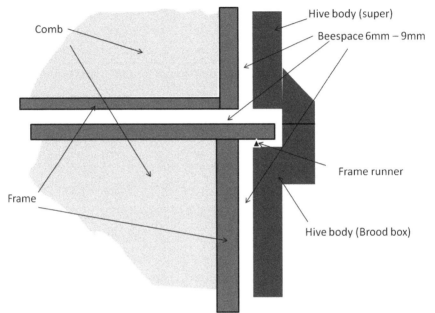

Comb

Hive body (super)

Beespace 6mm – 9mm

Frame runner

Frame

Hive body (Brood box)

Details of Bee Space inside National Hive

Bee space.

as the beespace. The beespace is 6 – 9mm across, a gap which allows the bees to move freely along it and around the frames. With a beespace this size, the gap remains clear and the frames do not become glued to the hive body with either wax or propolis, and so can be lifted individually from the hive easily. If a larger gap is used the bees build comb to bridge the gap and if a smaller gap is used the gap is sealed with propolis. Either way the frames quickly become firmly stuck to the hive body.

However this gap results in far more unfettered air movement than the bees normally have to cope with in a natural nest. It is obvious that honey bees do tolerate this and have done so since hives with moveable frames were first developed a hundred and fifty years ago. Whether this toleration has a cost to the bees and, if so, what that cost is, is not so clear. It is important that the hive is set up so that there are no through drafts, which reduce the control the colony will have over the temperature and humidity of its environment. Some authorities have suggested deliberately adding vents at the top of the hive to help the bees remove moisture as they process nectar during a flow. This creates a chimney affect that cools the hive in a way that bees cannot so easily control with their normal thermoregulation behaviour. Usually the bees eventually block these top vents with propolis. It is worthwhile observing that colonies that make a nest in a hollow tree usually make do with a single, remarkably small, entrance, often no more than 20 square centimetres in area.

It is often observed that honey bees in a moveable frame hive will nibble away the comb at the bottom and the lower parts of the sides creating the floating lower edges that occur in the natural honey bee nest. The reason why they do this is not clear. It could be to create additional passages for the movement of workers from one part of the hive to another, to alter the air flow within the hive or maybe to allow them to build queen cells hanging from the bottom edge of the comb. None of these reasons is totally convincing, but I do think that it indicates that the bees are expressing a small, if not major dissatisfaction, with the geometry and structure of the comb, held in frames, which they are being forced to use.

So it can be seen that honey bees that are put in man-made hives need to tolerate significantly different conditions than if they were living in a cavity in the wild. In addition to the different geometry within the hive, hives in an apiary usually have an entrance that is just a few inches above the ground. The air at that height, close to the ground, is likely to be far more humid and still than outside the entrance of a nest several metres high in a tree. Beekeepers can ameliorate this to some extent by ensuring that there is a space beneath a hive and the entrance is kept clear of weeds and grass so as to ensure good ventilation. They also need to consider whether it is appropriate to use a mesh floor to increase ventilation.

In many ways beekeepers try to design hives so that they imitate the natural cavity a bee would use. Traditionally hives were made from straw or wood, and they were constructed to have a volume that matched that of a natural cavity that honey bees would choose. For many centuries beekeepers used skeps, built from straw or wattle, and the bees would build comb as they would in the wild. But since the latter half of the nineteenth century beekeepers have increasingly used wooden hives that have within them moveable frames.

There are at least half a dozen moveable frames hives that are available from beekeeping suppliers in the UK, National, Langstroth, Commercial, Smith, Dadant, Rose etc. These are fundamentally the same in design and the advantages and disadvantages as far as the bees themselves are concerned are the same. The differences largely lie in their size and the choice that a beekeeper will make largely depends upon the environment in which the bees are being kept and the strain of bees being used. The more benign the climate and prolific the strain of bees, the larger the hive needs to be. The long and futile arguments that beekeepers engage in as to the best hive design serve no better purpose than to be an excuse for enjoying a drink in the pub after an apiary session. Another issue, which affects the beekeeper more than the bees, is the weight of the hive. Commercial bee farmers often work in teams and can employ mechanical lifting equipment and so that they can handle the larger, heavier hives. Hobby beekeepers generally work by themselves and may be more elderly and so lighter hives are more suitable. But that is a practical and economic issue, not an ethical choice affecting the wellbeing of the bees.

There is one other type of moveable frame hive that should also be considered separately and that is the double walled hive. The most well known of these is the WBC (named after William Broughton Carr). It is twice as expensive to buy as other standard hives but it is, with its pagoda type appearance, an attractive feature to have in the garden. Because it is double walled there is far more work involved in opening the hive and carrying out inspections of the bees. Also it is quite unsuitable for migratory beekeeping. The main issue is whether bees thrive in a WBC as opposed to a single wall hive. Though it may be supposed that a colony will be better insulated in the double walled hive, there appears to be no clear evidence about whether they do better or worse, one way or another.

The choice of material from which these hives are made may have an ethical component. Traditionally hives in the UK have been made from wood. The most popular timber used is western red cedar (Thuja plicata). This was originally sourced from forests high on the slopes of the Rocky Mountains in the northwest of the United States and in western Canada, where the trees had grown and developed slowly over a long period. Western red cedar is a conifer, not a true cedar. It can grow to be an extremely large tree, up to 70m in height and up to 4m in diameter. The tree is very long lived and many trees are known to have been alive for over a thousand years. The tree has now been naturalised in northern Europe, including the UK, and the eastern US. However, although it can flourish at lower altitudes, the timber produced in these places is not so fine grained as the native specimens. The timber is red-brown in colour when freshly sawn, straight grained with few knots. It can be rather brittle. When weathered it turns grey. It is a lightweight wood, about 25% lighter than other pines The wood contains a natural fungicide (thujaplicin) which inhibits rotting and this can remain effective for up to 100 years. The fungicide in the red cedar wood accumulates slowly as the tree is growing over many years and is hardly present in young trees. Hive manufacturers now source this timber from home grown forests, which prevents the depletion of the original ancient forests, but this timber is more coarse grained and probably contains less of the natural fungicide.

The alternative timber that can be used is pine, sometimes referred to as deal, which is cheaper and more available. Left untreated pine will begin to rot within a few years, but modern preservatives can make them equally long lasting, without being a risk to the bees. Hives made from pine are significantly heavier than those made from western red cedar.

In the last twenty years there has been a move to manufacturing hives from polystyrene. This has a number of advantages. They are a little cheaper to buy in the first place, they are lighter and they provide better insulation during the winter months. But there are disadvantages. The raw material they are manufactured from is oil, a non-renewable resource and they are more difficult to keep sterile. Wooden hives can be quickly sterilised by passing a blow torch flame across the inside. This is not possible with a polystyrene hive, and they need to be scrubbed with a solution of washing soda or something similar. And there is some debate as to whether polystyrene hives are really cheaper than wooden ones. A well made and maintained wooden hive can last over thirty years, possibly longer. The manufacturers of polystyrene hives claim that they should also last thirty years or more, but they are more easily damaged by hive tools during manipulations and can be destroyed totally by rodents or woodpeckers.

When frames are introduced into a hive, it is usual practice to fix sheets of foundation into the frames. Foundation is a sheet of bees wax, indented on both sides with the pattern of the hexagons the same size as the bees would normally produce when building comb. The hexagons indented on the foundation is usually sized for worker comb, but can be sized for drone brood. Foundation can be strengthened with diagonal or horizontal wiring melted into the wax sheets. By using foundation in frames, provided it is properly installed so it is in a plane and not bulging, it will guarantee that the bees will build their comb exactly where the beekeeper would wish.

Foundation is produced by several of the major beekeeping suppliers in the UK. Some of the wax that they use is produced by local beekeepers who are able to exchange blocks of wax for foundation. But the demand for foundation exceeds the supply of locally produced wax, so the suppliers must buy wax from overseas.

It is the source and quality of the wax used to produce foundation that raises the greatest concerns. Even wax produced in this country will be contaminated with the medications used to treat the bees for varroa. In addition it will contain traces of pathogens that are in the hive, such as the fungal spores of nosema and chalk brood. There are even greater concerns about the wax that is obtained from overseas, which may be sourced from countries where the prophylactic use of antibiotics is commonplace.

Most beekeepers accept that the risk to the bees from contaminated wax in foundation is very low. In order to produce the foundation the wax is melted by warming to a temperature that will kill most pathogens, and any contaminants which are still present become embedded in the wax when it solidifies. And besides, the foundation wax remains in the septum of the comb and is not usually ingested by the bees. The majority of the comb is built from fresh wax that the bees in the colony are producing and which is used to draw out the cell walls and cap the cells.

There are alternatives that can be used rather than manufactured foundation. Some beekeepers make their foundation using wax collected from their own beekeeping. Of course this could contain some contaminants but the beekeeper will have more control over what they might be. In addition it will save money. The disadvantage is that it is time consuming. Alternatively, the beekeeper may choose to not use foundation at all, in which case there is no guarantee that the bees will build the comb where required. If the comb is built so that goes diagonally across two or more frames, then it is impossible to inspect the bees as you would wish. As a compromise some beekeepers just use strips of foundation tacked to the top of the frame, which increases the probability of the comb being built correctly, but not entirely.

For a few years plastic foundation has been available, but amongst hobby beekeepers it has not been widely accepted. In theory it should be an ideal solution. It is cheap, it can be used repeatedly, it is strong and will not disintegrate in the extractor and the bees cannot nibble through it. The problem is that it is not that easy to persuade the bees to draw out the comb on the plastic foundation. It is recommended that the plastic foundation is coated with wax before being put in the hive. This can be done by dipping in molten wax, but this is an inconvenient procedure for the hobby beekeeper.

Ultimately the beekeeper has to make a choice between the convenience of using manufactured wax foundation or minimising the contamination that is introduced into the colony by using homemade foundation, starter strips or plastic foundation.

The queen excluder is usually regarded as an integral part of the modern moveable frame hive. It is effectively a bee sieve, a horizontal barrier splitting the brood area of the hive from the supers where the honey is stored. The queen excluder is designed so that the workers can move easily through it from one side to the other, but the queen and drones cannot, due to the greater width of their thorax.. By placing the queen excluder directly over the brood box, the queen, and therefore the brood, is excluded from the honey supers, which enables honey to be removed without affecting the brood and without having the honey contaminated with brood and the remains of the brood.

There are two different designs, a slotted steel or plastic square sheet or a wired grid excluder which is sometimes known as an Hertzog excluder. In some beekeeping circles, the slotted excluders are regarded as potentially damaging to the bees wings as they squeeze past the burred sharp edges of the slots as they move through the excluder. The wired grid is considered to be much more friendly for the bees, as the rounded wires are less likely to damage the workers wings, but they are considerably more expensive to purchase. Queen excluders should not be used during the winter months when the colony clusters around the honey stores. The winter cluster will not move through the hive if this would result in leaving behind the queen on the wrong side of a queen excluder, so, by leaving the queen excluder in place, it is possible that the colony can starve even though there are stores of honey in the hive.

Queen excluders must, to some extent, inhibit the natural flow of bees and air through the hive. There are beekeepers who choose not to use queen excluders at all, regarding them as unnecessary restrictions on the growth of the colony. The beekeeper must decide whether the convenience to the beekeeper and the lack of disruption to the colony when honey is removed, more than outweighs the negative effect on the bees.

To conclude, when we keep bees in the standard modern hive with moveable frames we are putting the bees, a feral creature, in an environment that to some extent must be alien. All the evidence is that the bees are able to adapt to it, but the difference between the nest that they would build in the wild and the hive they are given by the beekeeper is something beekeepers should always be aware of.

3 – Natural Beekeeping

In this chapter the issues are discussed which must be addressed when beekeepers choose to keep their honey bees in a way that more closely resembles how bees live in the wild.

Most beekeepers use hives with moveable frames, and with good reason, but there are alternative ways to keep bees. Just to reiterate, moveable frame hives allow for the easy removal of honey without harming the bees and allow beekeepers to easily and regularly inspect their colonies, checking for disease and controlling swarming.

Natural beekeepers will point out that there are several good reasons to keep bees besides the acquisition of honey. Firstly, honey bees are one of the most important pollinators, both of crops and the flora we see in the countryside, especially during the early spring. Anyone who starts keeping bees will have observed the immediate improvement in the yields from apple, pear and plum trees and soft fruit in their own garden. Uniquely amongst nectar and pollen collecting insects, honey bees are already functioning as fully developed colonies in the early spring. Throughout the winter they shelter in their hives as colonies consisting of thousands of adult worker bees and it just requires a warm day in March to release this massive pollinating force on to the local flora. I am writing this in the middle of January and only yesterday I did a mid winter inspection and some colonies, though not all, I estimate had about 25,000 bees. Compare this to a bumble bee. Bumble bees overwinter as mated queens, hibernating in small crevices in trees and walls. It requires the warmth of a March day to wake the queen from her winter hibernation and start the process of finding a suitable place for a nest, building it, laying eggs and raising brood. It may be two months later before the colony is starting to grow into a significant force of pollinators.

Another reason for keeping honey bees is simply for the pleasure of having them in your garden. To watch the bees busily flying in and out of their hives is relaxing and forms a connection between the watcher and the natural world.

During recent years there has been an increasing interest in natural beekeeping. Natural beekeepers aim to prioritise the wellbeing of the bees. This is achieved by designing hives which more closely meet the natural needs of the bees, reducing the number of occasions when the hive is opened, ensuring that apiaries are not overstocked with bees, minimising the use of manufactured foundation, not using queen excluders, not practising migratory beekeeping and avoiding, as much as possible, treating the bees with medications. Natural beekeepers maintain that this results in more placid and healthier bees.

These claims are understandable and possibly justified. It is when a hive is opened that bees are most likely to sting and when a bee stings it releases alarm pheromone, which heightens the aggressive, or if you prefer, defensive instincts of the bees. It is easy to envisage how this alarm pheromone lingers in the hive between inspections. All beekeepers will have observed that as the spring and summer progresses , the tetchiness of the bees tends to increase. There are probably several reasons for this, but repeated opening of the hive could be a contributory factor. Natural beekeepers aim to remove the stress on bees resulting from a lack of nutrition, movement and excessive interference and so it is not surprising if the bees are healthier if they are kept according to these aims.

However there are disadvantages. By not regularly inspecting the bees it is more difficult to control swarming. In an urban environment swarming can become a nuisance to neighbours and if bees are lost through swarming then the honey harvest will be minimal. Most significantly the natural beekeepers do not share to the same extent that exquisite experience and sense of wonder which comes from regularly inspecting the details of the organisation of life within the hive.

The natural beekeepers suggest that beekeeping has been distorted by the desire to collect from the honey bees as much honey as possible. This raises the interesting question as to whether honey should ever be collected, even if the cost to the wellbeing of the bees is minimised. In times past, honey was a valuable source of sugar, but now sucrose can be obtained cheaply and in vast quantities from sugar cane and sugar beat, to the extent that now sugar has become the scourge of the modern diet, contributing to the rising prevalence of obesity and type 2 diabetes. Though honey is mainly sugar, it is predominantly a mixture of fructose and glucose rather than the sucrose found in bags of granulated sugar and it contains flavours derived from nectars, traces of vitamins and antioxidants. It also has antibacterial properties. A demand for honey will always exist. If that demand is not met with good quality locally produced honey, it will be met from imported honey of a far more doubtful provenance.

It is relatively recently, that is since the introduction of moveable frame hives one hundred and fifty years ago, that beekeepers have concerned themselves with curbing the instinct of honey bees to swarm. Before that swarming was simply accepted as part of the process of managing honey bees. Moveable frame hives gave beekeepers the capability to control swarming. In an increasingly urbanised world, swarms are no longer accepted by the general public. We now have a large proportion of the population no longer at ease with natural events and they regard swarms as a nuisance and with fear. However if the bees are being kept in the countryside, swarming need not be regarded as being a problem. It can be argued that swarming does have benefits.

Swarming adds colonies to the population of feral honey bees and feral honey bees have an important role to play. Besides their role in pollination they are a source of drones needed to guarantee good mating of queens. The feral population also form an extended gene pool which is subjected to the pressures of natural selection and it is within the feral population that there is a likelihood of bees evolving that are adapted to the threats from new exotic pests and diseases. Feral colonies are under threat, long term and short term. The introduction of varroa, which requires the intervention of beekeepers to keep their bees alive, devastated the feral population of honey bees which were not recipients of the appropriate medications. In addition there are long term structural threats to feral honey bees. As the area of established woodland in the UK is reduced, there are fewer natural places for honey bees to form nests. England is already one of the least wooded countries in Europe. To their credit the government is trying to establish new areas of woodland across the midlands and in the north of England but it will be many decades before these are mature and the new forests include trees old enough to provide the type of cavities required by honey bees. In addition, modern building methods are less likely to provide accessible cavities, further depriving the feral honey bees of places to build their nests.

There are several hive types used by natural beekeepers which use top bars. One of the most popular of these is the Warré hive. This was developed in France by Abbe Warré in the 1930's. It was designed as an alternative to the moveable frame hives that were coming into vogue at that time and as a much cheaper alternative, enabling the ordinary working man to take up beekeeping. As with the standard hives, the Warré hives consist of a variable number of boxes placed on top of each other. But rather than the combs being contained within frames, the bees are encouraged to build combs that simply hang from bars stretched across the top of each box. As the hive expands new boxes are added at the bottom of the hive immediately above the floor. As the new comb is drawn in the lowest box, the queen follows, laying in the comb that has been newly drawn. The fact that brood is allowed to develop in new clean comb, contributes to the health and vigour of the colony. There are disadvantages. It is possible but not easy to inspect the

A view through the side window of a Warré hive.

brood on a regular basis and so swarm control is more difficult. There are other configurations of top bar hives. Some are built in long hives. There is also the sun hive promoted by the Natural Beekeeping Trust.

4 – Handling the Bees

It is when honeybees are lifted from their hives to be examined that they are most in jeopardy. This chapter discusses the risks, whether the bees should be subjected to these risks and how the risks should be mitigated.

Being sentient, the term mentioned in the EU Lisbon treaty, implies the ability to feel, perceive or to have subjective experiences. In particular, sentience implies the ability to feel pain. It does not, however, imply the ability to reason or to have an awareness of a future. Sentience is the least elaborate part of that ill-defined and not properly understood concept of consciousness. Whether insects truly are sentient as the term is generally understand is a good question, but our EU politicians tell us that they are, and whether scientists or beekeepers truly believe it or not it seems to me that it cannot be wrong to treat your bees as if they are.

There is the philosophical question as to whether honey bees are sentient as a colony, that is as a superorganism, as well as, or maybe instead of, individual bees. I guess most people would find this difficult to accept. The concept of honeybees as a superorganism was brought into prominence by the work of E O Wilson in the US and Jeugen Tautz in Germany. There are many respects in which a colony of honeybees, the superorganism, acts in the same way as a single living organism. It maintains a stable temperature, it defends itself as a single entity, it reproduces (that is it swarms) as single entity, it maintains reserves of nutrition as a single entity

Examining a frame of brood comb.

and it makes some decisions, such as selecting a new nest site, as a single entity.

A mammalian body is made up of tens of millions of living cells. In a similar way, a honey bee superorganism is made up of tens of thousands of individual worker bees. Every day a mammalian body casts off tens of thousands of cells, without causing pain or discomfort and replaces them at an equivalent rate. In a similar way the honey bee colony, in the middle of the summer, is losing between one and two thousand worker bees every day, which die through exhaustion, disease or by sacrificing themselves to defend the colony and each day the queen is laying a similar number of eggs to replenish these losses. To a colony of honey bees the defence of the colony is far more important than the life of a single worker, which is expendable. The sting of the honey bee is barbed. A set of internal levers push the sting deep into the skin of the victim before the venom is pumped in. When the bee tries to leave the site of the sting, the barbs prevent the removal of the sting and the sting, along with the associated organs are torn from the body of the bee. This leaves the bee fatally wounded and in a short while it dies. The sacrifice of the worker bee defending the colony is made without hesitation.

This still doesn't answer the question as to whether the honey bee superorganism feels pain. However there are occasions when a colony as a whole is under threat and then it acts and sounds as if it is distressed. For instance, I've seen colonies being culled because they have an infectious disease and the noise they made in their death throes was upsetting to my delicate sensibilities. To a lesser degree and to a less upsetting extent, most beekeepers will have seen how a colony reacts when the wellbeing of the whole colony is threatened such as when the hive is opened in cold or wintry weather.

It is interesting to examine the cognitive processes that bees go through when they are selecting a site for a swarm to settle to become a new colony. These are set out in detail in Thomas Seeley's book 'Honey Bee Democracy'. The details are beyond this tract, but the interesting thing is that the decision making process is fundamentally the same as what happens in a mammalian brain. If a honey bee superorganism is capable of cognitive decision making, the question must arise as to whether it is also capable of other cognitive processes such as memory or awareness of pain.

Regardless of how you regard the preceding paragraphs, there can be no good reason for beekeepers to not take care when they handle their bees. Inspection of the bees in hives usually involves the removal and viewing of individual frames on which there may be hundreds of bees. Beekeepers need to be aware that this is fraught with danger for the bees, as the process creates numerous opportunities to squash and injure them. It is not just for the altruistic motive of not wanting to hurt the bees that care needs to be taken. If a bee is squashed, this will release a cocktail of pheromones from the body of the bee, including the alarm pheromone, isopentyl acetate. This pheromone is released when a bee stings, attracting other bees to the spot where the pheromone was released and causing them to behave extremely defensively. The result is that if a bee is crushed, it is likely to initiate the defence instinct in the honey bee colony and increase the chance of being stung.

When handling bees most beekeepers wear gloves. The type of gloves needs some consideration. The beekeeping suppliers have available leather gloves with long cuffs that extend up to the elbow. If you are wearing these you are unlikely to be stung on the hands and fingers. However with thick gloves it is difficult to handle the frames and bees with sensitivity. And the thick gloves do not prevent the bees from stinging, they just prevent the sting penetrating through the leather to the skin of the beekeeper. The leather absorbs and retains the sting pheromone and this, as explained before, is a spur to further defensive behaviour. The alternative is to use bare hands or disposable gloves. Disposable gloves do not give much protection from the sting. If the bees wish to sting, the sting will penetrate the disposable glove easily. However it is probable that the sterile barrier they form between the bees and skin of the beekeeper does inhibit stinging behaviour. There are other advantages. Because they are disposable you can change gloves between hives or apiaries and so reduce the risk of the transmission of pathogens. When you handle bees you inevitably pick up propolis on your fingers, and so when you remove the disposable gloves at the end of an apiary visit you can drive home without putting sticky propolis on the steering wheel. However the most important advantage of using bare hands or disposable gloves is that you will become more sensitive to the feel and touch of the bees as you loosen and pick up the frames and, as a result, are less likely to harm the bees. There are beekeepers who , for medical reasons, cannot risk being stung on the fingers, and they have no choice and so must use leather gloves. Still it is a good idea for them to wear disposable gloves on top of the leather.

The smoker is not just an icon of beekeepers but an important tool. It has two functions. The first of these is to subdue the bees, reducing their inclination to sting. The theory behind this is that the smoke causes the bees to act in a precautionary way, making the preparations necessary should they need to abandon their hive in the face of a forest fire. They ingest as much honey as they can into their honey stomach so that they have resources available to set up a new nest. Having gorged themselves with honey they are unable to arch their abdomen, as they must do when they sting. Smoking does not reduce their defensive instincts, rather it temporarily makes them incapable of using their weapons. There is no unanimous view amongst beekeepers about the effective-ness of smoking as a method of calming bees. During most colony inspections I do not find it is necessary, but there are colonies and occasions when it is.

The second function, however, is essential for the wellbeing of the bees, and that is to drive bees away from areas in the hive where they may be harmed during colony inspections and manipula-tions. The first of these is when boxes are split apart. A little smoke should be used to drive the bees up into the box that is being removed so that the bees are not crushed when the box that is being removed is put down elsewhere. Secondly a little smoke should be used to drive the bees away from the queen excluder prior to moving it. Then a little smoke should be used to drive the bees away from the lugs of the frames before they are loosened and lifted from the hive. Finally a little smoke should be used to drive the bees away from the top of boxes when the hive is being reassembled once the manipulation is complete. In summary, whenever bees are at risk of being squashed or harmed it is good practice to drive them from the area where they are at risk.

Using water spray is the alternative method, which is both less effective and less intrusive on the life of the colony. When you open a hive of honey bees on a still day, you become aware of the smell of the hive, a complicated mix of odours from the wax, the honey and the pheromones that are being continually produced by the bees as a form of communication. That colony smell is an intrinsic part of the colony, and some authorities assign great value to it as a contributor to the colony wellbeing. The opening of the hive and the use of smoke must necessarily disrupt that colony odour during the time the hive is being inspected and for an extended period afterwards until the colony odour can be reestablished.

A metal hive tool is normally used to loosen the frames from the hive. The frames are positioned within the hive so that each frame is surrounded by a bee space. The bee space is between 6mm and 9mm. The Rev Langstroth discovered over a hundred and fifty years ago that the bees will maintain this space as a passage that the bees use to move about the hive. Nevertheless the frames must still have contact with the hive body where the lugs rest upon runners and at these points the bees use propolis to glue the frames to the hive. The hive tool is needed to lever the frames apart cracking the propolis seal. Care must be taken that there is no violent or sudden movement. And then before lifting the frame from the hive, the frame should be moved sideways away from the adjacent frame so that when it is lifted, the bees on the frame are not rubbed or rolled against those on the adjacent frame. All movements need to be smooth and fluent, ensuring that at no point is the frame shaken or jarred.

Many procedures in beekeeping require the queen to be located. This is not always that easy. The queen is 30% longer than a worker, her legs extend further beyond her body and the dorsal side of her thorax tends be less hairy, but in the colony during the summer the queen is just one out of 50,000 insects and so she is not always easy to spot.. To make finding the queen easier, many beekeepers mark their queens with a coloured spot on the back of the thorax. The colour used designates the year the queen was produced, white - years ending in 1 or 6, yellow - years ending in 2 or 7, red - years ending in 3 or 8, green - years ending in 4 or 9, blue - years ending in 5 or 0. The paints used are formu- lated so that they do not harm the queen. Nevertheless there are risks involved in the practice. To apply the paint spot requires that the queen is picked up or confined. Someone who is skilled can carry out the procedure safely, but all too often the queen can be damaged while being held or the paint spot is applied in a clumsy way and the paint runs on to the abdomen or elsewhere, blocking the spiracles through which the queen breathes. Crush- ing the abdomen can damage the delicate ovaries that lie within and reduce her egg laying potential.

Another procedure that is often performed is to clip one of the wings of the queen so that she cannot fly. This entails cutting off a third of the wing from one side with a pair of fine scissors. This delays the departure of swarms and so the time between the periodic colony inspections can be extended from seven days to nine days. Having to do fewer inspections is a benefit to the beekeeper and to the colony, but there is a potential cost. Beekeepers are assured that the queen feels no pain from this operation, but she can easily be damaged in other ways during the procedure if it is not done carefully, possibly losing an leg. Once her wing is clipped she unable to fly and if she should attempt to leave the colony with a swarm, she will fall to the ground and most likely perish. The swarm, without a queen, will return to its home colony.

5 – Nutrition

Apiaries, in which several colonies of bees are collected, are not natural, and the pressure on foraging resources which can result may find the honey bees struggling to find sufficient resources to guarantee their nutrition throughout the year. This chapter discusses the beekeepers responsibilities to keep his or her bees nourished.

Each year, if left to itself, a colony of honey bees will most likely swarm, at least once and in extreme cases up to five or six times. Each swarm that is produced will have the potential to form a new colony. It is obvious that when swarming occurs it is the exception rather than the rule if both the parent colony and the swarm survive through the winter until the following spring, otherwise, within a few years, the entire earth would be taken over by honey bees. There are natural predators of the honey bee, but the main limitation on the population of honey bees and the survival of a colony is the amount of forage or nutrition that is available.

Honey bees are social insects, but colonies of western honey bees, Apis mellifera, are not social superorganisms. In the wild, colonies of honey bees do not live together in groups as beekeepers force them to do in an apiary. On the contrary, the behaviour of Apis mellifera during the swarming process is designed to achieve as a wide distribution of colonies as possible. The reason for this behaviour is obvious as it minimises the competition for the finite amount of forage that is available. When a colony

*Even in the depth of winter, a hive will contain an active colony of bees,
each containing many thousands of insects.*

swarms, the scout bees look for a new home for the swarm that, on average, can be half a kilometre from the site of the parent colony, actively deselecting options which are close to the parent colony and which, by many other criteria, might be eminently suitable. Beekeepers often adopt the practice of setting up a bait hive in their apiary to attract swarms. Even when colonies within the apiary do swarm it is quite unusual for the bait hive to be taken over. On the occasions when a bait hive does attract a swarm, as often as not it is occupied by a swarm that originated from a neighbouring apiary.

Beekeepers need to understand that, by concentrating several honey bees colonies in an apiary, the bees may find they have difficulty in accessing sufficient nectar and pollen resources. There is a limit to the number of colonies that can be prudently maintained in a single apiary so as not to prejudice the wellbeing of the bees. The maximum viable number of colonies in a permanent apiary varies and will depend on the latitude, altitude, the geology and the amount and variety of flora in the vicinity of the apiary. In a rural apiary in England, I would suggest that eight colonies in a permanent apiary is a sensible maximum. This is no more than a rough guide. There will be situations in city centres, in areas of intensive agriculture or in uplands where even eight is too many. And there may be areas of mixed woodland and rough pasture in valley bottoms which may support more than eight. Here we are talking about the number of hives in permanent apiaries where the beekeeper is intending to keep his bees all year round and is relying on the local flora to provide sufficient forage to maintain the bees throughout the year. Temporary apiaries might be used by migratory beekeepers to take advantage of crops such as oil seed rape, borage or heather, and these could quite easily support a far greater number of colonies, but only for a short period.

Nectar is a solution of different sugars, predominantly sucrose and glucose, and is a source of carbohydrates for honey bees and many other insects. Carbohydrates are used by animals to provide them with energy. Honey bees may use the nectar immediately, or store it as honey. It is the ability to convert and then store nectar as honey that differentiates honey bees from almost all other nectar collecting insects. There are countless insects that obtain their energy nutrition from nectar. Because nectar is a weak solution of sugar, the sugar content rarely exceeding 40%, if left exposed to the atmosphere it will soon attract yeast spores from the air and will begin to ferment. Honey bees have evolved methods to reduce the water content of nectar to less than 20%, producing a supersaturated solution of sugars, mainly monosaccharides. This level of sugar concentration exceeds what can be tolerated by yeasts. To remove the water, the honey bees use a combination of chemical and physical processes. In addition to having evolved a method of reducing the water content, the honey bees have developed a method of storing honey in sealed wax cells so that the honey is not exposed to the moisture in the atmosphere. This is essential as honey is hygroscopic, meaning that it is a substance that tends to absorb moisture from the air. Absorbing moisture from the atmosphere increases the water content in the honey until it eventually reaches a level at which fermentation can occur. The coming together of the ability to collect nectar, the ability to reduce the water content of the nectar to make honey and the ability to store the honey so that it can be kept for indefinite periods is most remarkable and so it is not so unexpected that honey bees are one of the very few creatures to achieve this.

The ability to store honey allows the honey bee to continue to live as a social colony of thousands of individual insects during long periods when there is no nectar available, such as during the winter in temperate parts of Europe. The number of bees in a colony comes to a peak of between 50,000 and 60,000 at the end of June or beginning of July. For the next nine months this number gradually decreases to about 10,000 insects at the end of the following winter. For most of the winter months the bees form a cluster which gradually eats its way through the stored honey. Generally during the winter the bees will not leave the hive, but at around midday, on warm sunny days some bees will fly in order to defecate. Once the early spring flowers appear, such as snow-drops or crocuses, the bees will start to forage for short periods during the middle of the day, provided the weather is suitable.

During this winter period, the queen will spend her days at the centre of the cluster. Her rate of egg laying will be gradually reduced as autumn proceeds and the days shorten and for a short period at the end of November and during December she may cease to lay altogether. As climate change has caused winters to become warmer, the period when there is no brood being produced has become shorter and in the more southern parts of the UK it no longer exists. From the beginning of January, from a low base, the rate at which the queen lays will, over the next four months, gradually increase.

The energy that is released from the honey stores is used in two distinct ways: to fuel the metabolism of the adult bees and to nourish the growth of new brood. During the winter months the number of adult bees will gradually decrease, typically from 20,000 in October to 10,000 in March. As a result of these two factors, that is the energy needed to maintain adult metabolism which decreases as the winter passes, and the energy needed to raise brood, which increases after the winter solstice, the monthly requirement of energy remains fairly constant at about 2kg of honey for November, December, January, February and March. The total amount of honey stores required for an average sized colony for the seven months from October to mid April is variously estimated at between 18kg and 20kg.

During the summer months the energy requirement for a colony of bees is much greater, as both the population of adult bees and the number of brood being raised are much greater. It is estimated that during the month of July the energy required to fuel the colony is equivalent to over 25kg honey. This is equivalent to two supers full of honey. At that time of year the colony would usually be expected to collect an equivalent amount of nectar and, when the conditions are ideal, produce a surplus and so increase the honey stores in the hive. But it is obvious that if there is a prolonged period of inclement weather or there is no forage available, then a colony can very quickly exhaust the stores previously collected.

It is remarkable how quickly a large and apparently thriving colony of honey bees can succumb to starvation when there are no flowers available or the weather during the summer is inclement for an extended period, and so inhibits foraging, for several weeks. The larger the number of bees, both adult and brood, in a colony, the quicker a colony can starve. Once the levels of stores are reduced to a point where the colony is unable to feed the foragers enough honey to give them sufficient energy to fly, the colony loses heart, brood is abandoned and no longer produced and adult bees die. To start with, the bees that die within the colony are ejected from the entrance. But later the dead bees can just pile up on the floor of the hive. It is a depressing thing to find. In the late summer, when new queens should have become established egg layers and the chance of swarming is very low, there is temptation for beekeepers to leave their bees to enjoy the last few weeks of summer undisturbed. However it is still essential for the beekeeper to monitor the amount of stores that are in the hive. To do this there is no need to open up the hive fully. A sensible alternative is to heft or weigh the hive in order to check there are sufficient stores.

For many parts of this country, the periods in the spring and summer when honey bees actually make a surplus of honey are relatively short. Conditions are ideal when the weather is warm and calm and there is an abundance of nectar available. During these periods it is possible for the bees to fill a super of honey in less than a week. Visiting the apiary late in the evening after one of these days is a feast for the senses. You can hear the roar from the hives from about 20 metres as the bees spend their evening processing the nectar that was collected during the day, fanning the moisture extracted from the nectar out of the entrance of the hive. The air in the apiary is heavy with the scent of the nectar. These periods can occur in spring when the trees are in blossom, in the first half of May when the countryside is a patchwork of the yellow and golden fields of oil seed rape, in mid August on the moors of northern England when the heather is in bloom, in late August along the banks of rivers when acres of Himalayan balsam come into flower. These few weeks of surplus are required to see the bees through the periods off dearth and the long months of the winter. Beekeepers need to be aware, that in many areas and during many years there may be no natural net surplus of honey.

When you become a beekeeper you learn of something called the June gap. This is the point in the year when the blossoms of spring suddenly are no longer there and the countryside becomes a rather drab uniform green. Of course there are still some sources of nectar, but nowhere as near as much as there was during April and May. When an early warm spring is followed by a cool unsettled summer, the June gap can become extended.

In the past, before the moveable frame hive became commonplace, beekeepers would harvest their honey in the late summer by taking the entire honey stores from their bees. They would either kill the bees at the time or leave them in a such a condition that they were unlikely to survive the coming winter. Then, as now, beekeepers were not, in general, exploiting genuine surpluses, that is honey stores that the bees had accumulated over and beyond what they needed to survive the coming winter. Now, beekeepers try not to kill their bees and to obtain a honey harvest, adopt one of two alternative strategies. The first of these is migratory beekeeping, transporting the bees about the countryside so that they are in the right place at the right time to exploit the availability of large quantities of nectar from different sources, mainly monocultural crops such as oil seed rape which comes into blossom in the middle of April. Or secondly, they replace the honey that has been removed by feeding the colony with an equivalent amount of sugar in one form or another.

Migratory beekeeping is problematic. In the United States migratory beekeeping is a major industry, and without the pollination carried out by these itinerant beekeepers, large sectors of American agriculture would struggle to be viable. At the centre of the migratory cycle are the almond orchards of California. Many bee farmers make half their income from pollination fees from the almonds. After the almond blossom some beekeepers take their honeybees to cherry, plum and avocado orchards in California and apple and cherry orchards in Washington State. During the summer months, the beekeepers head east to fields of alfalfa, sunflowers and clover in North and South Dakota, where the bees produce the bulk of their honey for the year, while others visit squashes in Texas, clementines and tangerines in Florida, cranberries in Wisconsin and blueberries in Michigan and Maine. The bees are transported from one location to another packed onto large lorries, each carrying hundreds of hives.

The stresses that are placed on the bees by these journeys have resulted in large and consistent losses of colonies, with beekeepers in the US losing up to 30% of their colonies every year. CCD (Colony Collapse Disorder) was thought to be at least partially due to the stresses that are imposed upon the bees by these practices.

Taking advantage of the bonanza from oil seed rape.

In the UK, there are far fewer bee farmers that move their colonies around the country in this sort of way, and the average number of colonies that are managed by each bee farmer is much smaller. In the UK the majority of colonies kept by beekeepers belong to hobby beekeepers who have less than 20 colonies. Nevertheless the migration of bees has had, in the past, a serious deleterious effect on beekeeping in the UK, accelerating the spread of diseases, and in particular varroa, throughout the UK. Small hive beetle is not currently in the UK. When it does arrive, as it almost definitely will in the future, there is a high probability that the transmission of this pest will be accelerated by the migratory activities of beekeepers.

Migratory beekeeping provides a living for people, and enables large quantities of honey to be produced. The people who carry out migratory beekeeping are in the main good beekeepers and look after their bees in a conscientious way. It is not just large bee farmers who move bees. For instance, when hobby beekeepers move home they may take their bees with them. There is a countrywide trade in queen bees, and when queens are moved they are accompanied by a small number of workers any of which could be carrying pests or diseases. In the spring many associations hold auctions where bees are sold and bought and then moved to a different part of the country.

It is almost inevitable that if bees are kept in hives, then they will be moved. Most hives are designed so that the bees can be moved. And every time bees are moved there is a risk that diseases and pests of the honey bee can be spread. There are advantages in being able to move bees for purely management reasons. Hobby beekeepers often find that having two apiaries can be useful. For instance if you need to unite colonies, this can be difficult in the confines of a single apiary. Before colonies can be united they first need to be brought together so they are within a couple of feet of each other. This is not so easy within an apiary as there is a working rule that hives can be moved less than two feet or more than two miles. If this rule is broken, then the flying bees return to the original site, only to find their hive has disappeared. If a beekeeper has more than one apiary this problem can be solved by moving one hive to a second apiary and then three or four weeks later bringing it back to the original apiary but now placing it next to the hive to which it is to be united.

Beekeepers feel that they have the right to take the honey when it is there to take. The honey is regarded as a rent that is due to them for providing the bees with a home and a place to live. In taking the honey it needs to be born in mind that it is very probable that the beekeeper is taking from them the stores they need to survive for the rest of the year. It is the beekeeper's responsibility to ensure that their colonies are provided with sufficient stores. Beekeepers maintain that the honey produced from collected nectar is superior to honey that is manufactured by the bees when fed sugar syrup. Undoubtedly as a honey crop to be sold to paying customers that is true, and in fact it is verging on the fraudulent to sell honey produced by feeding bees sugar solution or corn syrup.

However it is more debatable if natural honey from blossom nectar better enables a honey bee colony to survive the winter better than an equivalent amount of honey produced from a solution of pure sugar. It is true that natural honey contains greater amounts of proteins, vitamins and trace elements than honey from pure sugar. Bees do need these honey additives, but, regardless of whether honey is removed, they are all present in the hive in the pollen that remains once the honey is taken or not. On the other hand some natural honeys granulate and by winter can be very difficult for the bees to ingest. When the stored honey has granulated, the bees need to collect water to help dissolve the crystals. The water is usually available but in the middle of winter it often can be too cold for the bees to fly and even if there are mild spells it requires a significant expenditure of energy to collect the water.

During September, whether honey has been removed or not, the beekeeper must ensure that each colony has sufficient stores to sustain itself until the following April. A good guideline figure for the stores required at the end of September is 18kg to 20kg, though, to some extent, the amount will depend upon the size of the colony. The sugar is usually fed as concentrated sugar solution, two parts of sugar to one part of water by weight. If the full requirement of honey stores is to come from sugar solution fed by the beekeeper, then at least 18 litres of concentrated sugar solution would be required. That is quite a lot and is heavy to handle. When feeding, it is important to use the type of feeder that is appropriate for the amount being fed. Though it is possible to supplement the stores by feeding during the winter, generally it is better policy to fully feed the colony in the autumn so that it can survive the winter without further intervention. If additional sugar is required during the course of the winter then it is usual to feed with fondant rather than sugar solution.

A second component of the nutrition required by the bees comes from pollen. Pollen provides proteins, vitamins and trace elements. Proteins are essential to enable the larvae to grow and develop. The adult bees also require small amount of proteins to replace tissue and to synthesise various pheromones, hormones and enzymes. Wax is largely synthesised from carbohydrates but a small amount of protein is also needed. Pollen is essential to the well being of the colony but it is not usually so mission critical to the bees, at least in the UK. The amount of pollen required is much less than the amount of nectar and is available in most areas almost throughout the year. The availability of pollen is increased as the bees are able to collect pollen from plants that do not produce nectar, such as hazel. The bees store pollen in the hive, usually in an area close to the brood nest, but in relatively small amounts compared to honey. If pollen is temporarily not available the colony can reduce the amount of brood they are producing, without prejudicing the long term wellbeing of the colony.
A factor that does appear to affect the health of the colony is whether the colony is able to access a variety of pollen sources. Different pollens contain slightly different mixes of amino acids. The bees seem to be able to detect that some pollens are better than others and will go out of their way to ensure that a variety of pollens are being foraged.

It is not normally necessary for the beekeeper to give pollen supplements to colonies of honey bees. Pollen supplement prod-ucts are available from beekeeping suppliers but their availability does not necessarily imply their necessity.

Besides nectar and pollen, the forager bees can also be required to collect water. Water is required to dilute honey stores so that they in a form in which they can be directly used by the bees. Water is also required to cool the hive during warm weather. For some of the year there is no necessity to collect water. If there is a nectar flow the bees can use the nectar for nutrition rather than the stores. During the winter the bees use the condensate that collects on the walls of the hives to dilute honey.

The main period when water is required is during early spring when the bees are increasing the raising of brood and are relying on stores for the nutrition. In our climate a lack of water is not normal thought to be an issue. On the other hand water is a heavy commodity and in some areas, on chalk lands for instance, there may be no lying water. By providing a shallow tray of water in the apiary the beekeeper can save the colony considerable expenditure of energy. It also needs to be born in mind that honey bees seeking water can be a nuisance to neighbours, using swimming pools and garden ponds as water sources. If the beekeeper does provide a water source it is preferable to have a gently sloping edge leading into the water.

This chapter seems to have been largely about honey bee nutrition rather than ethics. But it is in this aspect of beekeeping that beekeepers most often fail their bees and as often as not it is because of poor understanding of the nutritional needs of their bees. Whether by overstocking apiaries, by failing to provide them with sufficient stores to get through the winter or by failing to recognise a period of dearth during the summer, a lack of nutrition can lead to a colony's demise. The fact is that these losses of honey bees are preventable and it is unethical to keep bees and fail to ensure they have sufficient nutrition.

6 – Diseases of the Honey Bees

When beekeepers collect colonies of bees together in an apiary, they can alter the balance between the bees and the diseases to which they may be subject. This chapter discusses the implications of disease and the control of disease on the work of the beekeeper.

Most diseases have evolved, over millions of years, to live on a specific host and usually a disease, whether it is a virus, bacteria, fungi or mite will have adapted to survive in balance with its host. If it is so virulent that it kills its host then that will lead to its own demise. The most dangerous diseases we encounter are usually those that have crossed from one species to another or moved from one part of the globe to another so that balance has not had sufficient time to become established. Both of these causes are usually the result of human intervention rather than natural causes. As beekeepers, we can very easily alter that balance between our bees and the diseases that they may be subjected to.

If a species is widely distributed in nature, any disease that infects that species will tend to be highly infectious. Honey bee colonies, as we have said, tend to be widely spaced in the wild, and so the diseases that infect them, such as European Foul Brood, have, of necessity, evolved to become highly infectious. When a beekeeper brings a number of colonies together in an apiary, this is one way in which the natural balance between honey bees and their diseases can be altered.

Honey bees have a number of physical and behavioural characteristics that give them a natural resistance to disease. Their bodies are
encased in an exoskeleton which is impervious to bacteria, viruses
and the spores of fungi. They make propolis from resin produced
by some plants and collected by the bees, and this acts as a natural
antibiotic. The propolis is used to line the inner surfaces of their
hives, to reduce the size of the hive entrance if necessary, to block
up unwanted gaps in the hive, to mummify small mammals that
may have died within the hive and to line the inner surfaces of
brood cells between each cycle of brood. In addition they have a
number of hygienic behaviours. These include removing bees that
have died within the hive and recognising and discarding larvae
that are diseased. They also have an innate immune system that
will give them protection from some, but not all, viral diseases.
This is an immunity that they inherit through their genome rather
than an adaptive immune system which is able to adapt to counter
viral or bacterial threats that they have not previously encountered.

Honey bees, as is the case with all insects, do not have the adaptive immune system which mammals have. An adaptive immune
system learns, in a relatively short period of time, to be able to
fight off infections which it has not previously come into contact
with. Without an adaptive immune system mammals would struggle to survive as their lungs, by absorbing oxygen into their blood,
which is then circulated throughout their body, are very vulnerable to the absorption of bacteria and viruses carried in the atmosphere. Insects, relying upon a network of small tubes, trachea and
tracheoles, to bring oxygen into their bodies, do not have lungs or
use blood to carry oxygen around their bodies and so do not have

this vulnerability. Insects do not have blood that circulates around the body distributing oxygen, but they have an equivalent system that circulates haemolymph, distributing nutrition and hormones. Once a potentially deadly virus enters the haemolymph, having bye-passed their exoskeleton, unless they have an innate immunity to the virus, they will quickly succumb to it.

It is this vulnerability that has made mites that prey on honey bees so deleterious to their wellbeing. During the last hundred years there have been two mites that caused immense losses to the honey bee population in the UK and elsewhere in the world. The first of these was acarine, which almost wiped out the native British bee in the first couple of decades of the twentieth century. The acarine mite is too small to be seen with the naked eye and largely lives in the thoracic trachea of the adult honey bee. It requires a dissecting microscope to see it. The second of these mites is varroa. This mite was originally a parasite of Apis cerana, a species of honey bee native in the far east of Asia. Probably due to human activity it transferred to Apis mellifera and the infestation gradually moved west across central Asia and Europe, being first found in England in 1992. Varroa is a much larger mite than acarine, about the size of a pinhead, and easily seen with the naked eye. It breeds in the brood cells of the honey bee, feeding on the pupae and then has a phoretic stage when it lives on the body of adult bees. Both mites feed on the haemolymph of the honey bee, and to gain access to the haemolymph they are equipped with a proboscis that is able to pierce through the exoskeleton. By puncturing the exoskeleton the mites allow viruses to bye-pass this important defence and it has been the viral diseases that are introduced in this way that have proved so devastating to the honey bee population.

The spread of varroa is a salutary lesson as to how damaging the activity of beekeepers can be to the wellbeing of honey bees. Varroa was first found in Devon, England in 1992. Sixteen years later it had reached Nairn in the north of Scotland, having spread the length of England and across most of Scotland, a distance of over 600 miles, averaging about 40 miles each year. This might not seem a very fast rate of dispersal until it is considered that varroa cannot fly. In nature it depends for its transmission from one colony to another on the drifting of bees, robbing and swarming during its phoretic stage. It is difficult to envisage how these methods of transmission would result in a rate of transmission of more than five miles per year. At that rate, twenty-five years after varroa first reached Devon in the south west of England the varroa mite would only just have reached Cheltenham. But of course we have beekeepers who collect swarms, buy and sell colonies and take part in migratory beekeeping. Even when the government imposed travel restrictions on bees, beekeepers from the south of England were bringing colonies to the North Yorkshire moors to take advantage of the heather blossom.

But that was in the past. It happened and now we must live with the situation that exists. Varroa, once it is in a colony, and if there is no intervention, will, in the vast majority of cases, eventually kill the colony. Slowly the varroa population increases, doubling its population each month of the spring and summer. Three years after a single varroa mite is introduced into a hive, if there is no intervention, the number of mites, and the viral diseases that accompany them, will overwhelm the colony.

Of course there have been interventions. The first of these treatments was developed by a couple of the world's leading manufacturers of pesticides used in agriculture. They were based on pyrethroids. To start with it seemed to be a magic bullet and had an efficacy of 99%. But after a few years it was found that it was no longer being as effective, as the mites had developed a resistance to the medication. For a couple of years, in the middle of the first decade of the twenty-first century, beekeepers had enormous winter losses of colonies of honey bees. Fortunately there was another answer, using a concept employed in horticulture called integrated pest management, where the levels of the pest are kept below those that would cause the colony's wellbeing to be adversely affected by using a variety of treatments, some chemical, some biotechnical. The newer chemicals do not have the high rates of efficacy of the original treatments based on pyrethroids, and so must be used in combination with other treatments. If beekeepers want to keep varroa levels in their colonies at a low level, they have a tool box that works, but it does require disciplined application on the part of the beekeeper.

The government Bee Unit and BBKA advise that honey bee colonies will inevitably succumb to varroa if the colonies are not treated, but this is not universally accepted. There have been, and continue to be, attempts to breed honey bees that can fight against varroa infestation and some beekeepers claim that they are no longer treating their bees for varroa, their bees are not succumbing to the mite and the levels of the mite within their hives are

Treating for varroa in the mid winter using oxalic acid.

remaining stable and sustainable. On the other side it is argued that beekeepers who do not treat their bees are allowing the varroa levels in their colonies to rise, and this not only affects the wellbeing of their own colonies but also will effect the colonies of neighbouring beekeepers. This is a debate that will continue.

In the long term it might be expected that both the honey bee and the varroa mite will evolve so that the host and the pest can live in balance. The honey bee could develop hygienic behaviours that successfully reduce the varroa population within the hive. The varroa mite could evolve so it becomes less virulent. There is no clear evidence that this is happening yet.

There are other diseases that beekeepers need to take precautions against. The most important of these is are the foul brood diseases, European Foul Brood (EFB) and American Foul Brood (AFB). Both of these diseases are very infectious and difficult to eliminate once they become established in a colony. Both effect the bees in the brood stage. They differ in that EFB affects the brood while it is in the larval stage and AFB affects the brood in the pupal stage. Both are notifiable diseases and must be reported to the bee inspectors belonging to the National Bee Unit who take on the responsibility of bringing the outbreak under control. The main responsibility of the beekeeper is to learn to recognise these diseases when they occur and to inform the bee inspectors if an outbreak is detected. Each beekeeper has the responsibility to keep their bees as healthy as possible, not just for the sake of the wellbeing of their own bees but for the sake of those that belong to neighbouring beekeepers, and indeed for the wellbeing of the feral colonies in the area.

In the wild, a colony in a hollow tree or other natural cavity will not usually survive indefinitely. Even though every one or two years the queen leading the colony will have been replaced this is not sufficient to guarantee the indefinite survival of the colony. Over a longer period of time the combs in the cavity will become older and the individual cells will become lined with an increasing thickness of propolis. The comb becomes a reservoir for disease, and the colony will be weakened and eventually die out. Once the bees are gone, wax moths will colonise the cavity and quickly clean out the wax and propolis. When this is completed they too will abandon the cavity, leaving a pristine space, ideal for a fresh swarm looking for a place to set up home.

A similar thing happens to comb in hives managed by beekeepers. Over a single year, the combs at the centre of the brood nest can be used and reused up to eight times to raise brood. Over the years the brood comb becomes older, a reservoir for disease and increasingly damaged so that less of it is available for the queen to lay in. Rather than allowing a slow decline, the beekeeper has the option of periodically replacing the old brood comb with new foundation, every one to two years.

Exotic diseases.

Beekeeping in the UK is facing a number of potentially severe threats, which could have serious implications for beekeepers in the next few years. The first of these is the Asian Hornet. The Asian Hornet preys on nectar feeding insects and in particular honey bees, hawking in front of hives and killing foraging workers which are returning to the hive. These are carried back to the hornet's nest to feed their larvae. Though this doesn't necessarily kill the honey bee colony, it can seriously weaken it, reducing the amount of honey it is likely to collect and prejudicing its ability to survive the winter . The Asian Hornet usually nests high up in trees and the nest is difficult to find. In 2018, at the time of writing, the hornet is well established on the other side of the English Channel in northern France and there have been a couple of isolated cases of the hornet being seen in the south west of England. In both these cases the nest was found and destroyed before mated queens emerged.

There is an inevitability that sooner or later a colony of the asian hornet will complete its full summer cycle and produce a significant number of mated queens. When that happens it will be very difficult to control in the following years. This is what happened in France. There is a forlorn hope that the cooler climate in the UK will prevent the asian hornet thriving and becoming established. It is hoped that through the vigilance of beekeepers it may be possible to significantly delay the time when the asian hornet becomes established in the UK. It is possible to lure the hornets into traps. If traps are set up in the spring then there is a chance that queens of hornets can be caught before they can establish a nest and start to lay. The down side to this strategy is that the traps will also catch the queens of native hornets, which are not a threat to bees, and queen wasps. I'm sure many might suggest that the less wasps the better. However wasps are an important predator of aphids during the majority of the summer and so important to gardeners. It is only at the end of the summer they become a nuisance. So spring trapping of queens needs to be carried out selectively.

The second threat is the small hive beetle. This pest originated in Africa and has caused serious damage to beekeeping in Australia and North America. There have been a number of incidents where the beetle has been found in Portugal and southern Italy. The outbreak in Italy has not definitely been controlled. At the present time there have been no incidents of the pest being found in the UK. The female beetle lays its eggs in the hive of the honey bee and it spends its larval stage in the hive doing significant damage to the honey comb. After about two weeks the larvae are ready to pupate and leave the hive and burrow into the soil near the hive. The pupal stage lasts about three or four weeks. When the adult beetles emerge they mate and fly to find a hive in which they can breed. Hive beetles may have four to five generations a year during the warmer seasons.

The threat of the small hive beetle is less imminent in the UK. The small hive beetle is very vulnerable during its pupal stage and nematodes have been found that can be sprayed on the soil around infected hives which will destroy the beetle while it is in its pupal stage.

For both of these exotic pests the main defence is the vigilance of the beekeeping fraternity. Beekeepers have an obligation to familiarise themselves with these pests and then put into place the procedures and disciplines to look for them and report them to the authorities. The small hive beetle is a notifiable disease. The asian hornet is not strictly speaking a notifiable disease but nevertheless it should be reported to the Bee Unit.

This book is not about bee diseases and there are many other diseases I have not mentioned. The main point that I hope is being made is the ethical necessity for every beekeeper to keep his or her bees in such a way as to minimise the chances of diseases, to be familiar with the diseases off the honey bee, be vigilant in the lookout for disease and then take whatever steps are necessary to bring an outbreak of disease under control.

7 – Breeding of Honey Bees

This chapter discusses whether improvements bred into honey bees by beekeepers may have long term deleterious effects on the ability of the honey bees to survive.

A gene, a unit of heredity, is a string of DNA situated at a fixed position on a chromosome. Each of the thousands of characteristics that define a living species is defined by a gene or a set of genes. Within a species, there can be many different forms of any given gene. These different forms of genes are called alleles. New alleles are continually being formed through random mutation. Most of these variations are short lived as they do not lead to viable phenotypes. When it is realised that there are hundreds of thousands of genes and each of these genes can take many different forms, it is not so difficult to understand why almost all individuals within a species are unique, both in their genotype and consequently in their phenotype.

A century and a half ago, Charles Darwin had the fundamental insight that transformed our understanding of the development of the diversity of life on earth. He observed that within a species there can be countless possible variations of the phenotype. He realised that variation within a species leads to a natural selection, so that an individual with the set of alleles (though he didn't use that term) that makes the resulting phenotype best suited to its niche in the environment is the most likely to survive and procre-

Removing ripe queen cells from a cell bar prior to placing them in a mating nucleus.

ate. As the environments change, species are able to adapt and make themselves more competitive in any new environment.

This process is at the simplest level of evolution and when combined with the existence of geographical barriers such as seas or mountain ranges it is responsible for producing the different strains or subspecies.

When humans become involved with breeding, their choices of what is preferable become the drivers of selection rather than natural selection, through survival of the fittest, as occurs in the wild.

Any beekeepers who keep several colonies over a period of time cannot but be aware that colonies of bees are not all the same, some are better than others. There are a number of characteristics which vary and these include profligacy, the ability to produce surpluses of honey and temper.

All domestic animals kept by man have been subject to selective breeding. In most cases the breeder can control both the male and female lines. The mating of the honey bee takes place in the open air and at a height of ten metres or so above the ground. It has never been possible to induce honey bee mating in other situations such as in an enclosed space. In a breeding program for the honey bee it is easy to select on the female side but only possible to control the male line by either using instrumental insemination techniques or by using isolated mating apiaries. These options are not normally available to the small time beekeepers that make up the majority of beekeepers in the UK. Selective breeding on the male line is made even more difficult as the honey bee in polyandrous, that is the queen mates with many drones, typically between ten and twenty. Many beekeepers try to improve their bees through breeding, but selecting only on the female line.

Another fundamental difference between the breeding programs for most domesticated animals and honey bees is that the resulting offspring from selective breeding of domesticated animals are kept separate from their cousins who remain living in the wild. Most domestic dogs, cattle, sheep and pigs would not be able to survive if they had to live in the wild and could not compete with their cousins bred in the wild. Their human breeders, having bred in them the characteristics that suit their purposes, must then protect them from the harsh demands of the wild. We may keep our bees in wooden hives in the garden, monitoring their health and supplementing their stores as necessary , but in many other ways our bees remain a part of the feral population of honey bees. They must compete with feral bees for forage, they mate with drones that may have originated from the feral population, they must defend their honey stores from being robbed by colonies of feral honey bees and other insects such as wasps that live in the wild and they will be subjected to the same pests and diseases as the feral population. When selectively breeding cattle, it is not necessary for the farmer to be concerned whether the resulting progeny will be able to defend themselves against their natural predators such as wolves or whether they will be able to compete with herds of wild cattle, as the farmer will provide the necessary protection. A similar consideration does not apply to those who breed honey bees.

The characteristics that is most often subject to selection in a honey bee breeding program is temper. Bees that are overly defensive are unpleasant to work with and this can lead to a reluctance on the part of the beekeeper to carry out the management tasks that are necessary, such as swarm control and health checks.

When considering the honey bees ability to sting, it is tempting to believe that the honey bee's sting evolved in order to fend off human beings and in particular beekeepers. Honey bees, through their industry and skill, are able to collect and store large quantities of honey. This is a product that is not just of value to the bees, but also to many other mammals, birds and insects. Human beings have been added to the list of possible predators of honey bees relatively recently. The honey bee evolved into something like its current form about 40 million years ago and for the last 40 million years has remained in a state of stasis. 40 million years ago was at the end of a period when flowering plants were evolving which required insect pollination. Human beings evolved into their current form about two million years ago. For most of the time that honey bees have existed their main predators have been other colonies of honey bees and closely related insects such as wasps, and that remains the case today. The tendency of bees to sting beekeepers is no more than an unintended consequence of their evolved ability to defend themselves from other insects and animals.

It is therefore necessary to ask whether selecting bees that are less defensive by nature is likely to result in bees that are less able to defend themselves from their main predators.

Another characteristic that some beekeepers try to breed out is excessive collection of propolis. Propolis is collected by bees as an exudate from botanical sources, in particular tree buds and sap flows, and when mixed with beeswax and saliva, forms a resinous substance. It is used by the bees to seal unwanted gaps in the hive, to reduce the entrance size, to strengthen the edge of cells and the anchorage of the combs to the hive body and as a antibiotic lining to brood cells. It can be collected by beekeepers and used to make antibiotic gels and tinctures. However if a colony of honey bees collects too much propolis, it forms a sticky gum on the fingers of beekeepers and hive tool and makes the frames more difficult to remove. As a result some beekeepers have tried, through selection, to breed out the tendency to collect excessive amounts of propolis. A question that needs to be asked is whether it is better for the wellbeing of the bees to reduce the amount of this natural antibiotic being brought into the hive.

Selective breeding involves the removal of an allele defining a characteristic, replacing it with an alternative allele. The ethical question that is posed is whether the benefit to the breeder is more or less outweighed by a detriment to the wellbeing of the bees.

I previously referred to instrumental insemination techniques, which enable the breeders to control the male side of the breeding selection as well as the female. These techniques are well established and used in research institutions and by large breeding programs. A number of small time beekeepers also use the technique. But it requires a financial investment and the attainment of skills that are beyond most hobby beekeepers.

The semen needs to be collected from quite a large number of drones for each insemination. The collection procedure results in the death of each drone used, but as drones die when they mate naturally this should not be regarded with too much distaste. The virgin queens are anaesthetised, normally with carbon dioxide and secured in a tube so that the rear of their abdomen protrudes. The bursa copulatrix leading to the vagina is held open with small hooks and a pipette containing the semen is inserted, so that the semen can be pumped in. The tube holding the queen, the pipette and the hooks are controlled by an operator using a binocular microscope. There is no evidence that the queen suffers and she recovers quickly from being anaesthetised.

It has been shown that it is possible to achieve a certain degree of improvement of some characteristics of honey bees by only selecting on the maternal side. But if we want to produce bees that have all characteristics that we desire, it is necessary to select using both male and female parents and if breeders are to control the male line, instrumental insemination is a necessity. To many beekeepers the whole process seems cruel and unnecessary. The quality of the queens obtained is often not that satisfactory, there is a risk that the queens can be harmed by the insemination process and the repeated use of instrumental insemination may lead to inbreeding and loss of diversity.

As mentioned before the queen is polyandrous so that we can see that the natural mating of honey bees has evolved so that within a colony there is a diversity of genomes. The aim of instrumental insemination is to produce a uniformity of bees conforming to the characteristics desired by the breeder. So we have a situation where the aims of instrumental insemination run counter to results of many millions of years of evolution. There will always be some debate as to whether it is really a good idea and whether it is likely to have a long term deleterious impact on the wellbeing of the bees.

8 – Other Beekeepers

In this chapter there is a discussion on the beekeepers relationship with other beekeepers. In particular there is a discussion about the contribution of beekeeping associations in the UK and how the ordinary beekeeper can both benefit and be involved.

Once you become a beekeeper you join a large group of like-minded people. As there is no mandatory registration of beekeepers in England it is not possible to say with any certainty how many beekeepers there actually are in this country, but for the last few years there have been about 25,000 full members of the British Beekeepers Association (BBKA) which serves England. In addition there is a Bee Farmers Association. There are likely to be a similar number of beekeepers who are not members of either association. It should be said that the other countries within the UK have similar organisations that correspond to the BBKA, as indeed do many countries throughout the world

Beekeeping is largely a solitary activity and therefore it is not altogether surprising that it attracts people who do not choose to be in an association and prefer not to make their beekeeping into a social activity. But even for those who are of a solitary disposition, being a member of an association brings with it quite a few benefits. Most associations often organise bulk buying deals, there can be equipment to hire or borrow and many associations have libraries. By joining your local village, town or city association, in most cases, you join, by affiliation, the county association and the BBKA. These also confer benefits on the members. For example the BBKA membership includes a third party liability insurance. However in many peoples eyes, the most important benefit is training and education. Beekeeping requires knowledge and understanding. It is true that this can be obtained to some extent through experience or by private study. There are thousands of books that have been written about bees and beekeeping, and remarkably there are booksellers that are devoted solely to selling books about bees. But by far the most satisfactory method of learning is through the courses and talks organised by the hundreds of beekeepers associations that are scattered across the country. These have reflected the changing circumstances and challenges that beekeepers are encountering over the years, in a way that books cannot, and have provided a forum for debate on the issues that are facing beekeeping.

Twenty five years ago the arrival of varroa into the UK led, over the decade that followed, to a complete reappraisal of beekeeping methods. Eventually the concept of integrated pest management was introduced into beekeeping circles and this provided the basis for being able to live with this pest. In the next few years there must be high likelihood that the Asian hornet and small hive beetle will become established in this country. Naturally the prospects of this happening is a reason for some despondency amongst beekeepers. Looking back, there is much to be learnt from the actions taken to cope with the arrival of varroa. When it first arrived in the UK in 1992 there were similar forebodings, but beekeepers experimented with new methods and treatments and the knowledge of those methods that proved successful were disseminated by the beekeeping associations. The overall result was that the health of bees improved, winter losses were reduced, honey harvests improved and the number of people taking up beekeeping increased. It wasn't all good news. Some beekeepers failed to heed the warnings and advice, lost their bees and gave up the craft entirely.

Whether the combined resources and resourcefulness of the beekeepers of England will find methods to deal with the problems created with the probable arrival of the Asian hornet or the small hive beetle only the future will tell, but history gives us some cause for optimism. If there is to be a solution, as with varroa, it will depend on the active participation of the beekeeping associations. Of course the research departments of government and the universities must establish the science, the natural history, the genetics and the methodologies that can be used to combat these threats, but it is the associations that will disseminate the information, persuade and cajole their members to alter their beekeeping practices to deal with the new risks, and give feedback as to which approach is the most effective.

Besides the response that the associations have to new situations, most associations are actively involved in providing training for people who are just beginning their beekeeping and for those who are wishing to progress to more advanced skills. As often as not the beginners courses are split into two parts, the theory and the practical. The theory is necessary, because, without the knowledge of the life cycle and natural history of the honey bee, it is not possible

An apiary training meeting.

to understand and appreciate the practical procedures which are a standard part of beekeeping.

County and the national associations usually take on the responsibility of teaching the more advanced aspects of beekeeping and preparing beekeepers to take the higher level examinations. The BBKA incorporates an examination board that organises examinations at all the different levels, from basic beekeeping to advanced. More of that later.

I'm writing here about ethics and ethics is about behaviour and how it effects others and whether that behaviour is right and wrong. Doing something that is purely for your own benefit is hardly a question of ethics. But being a member of a beekeepers association is not simply a matter of self interest. If there are to be those that are learning, there must be those that do the teaching. And in addition once there is an association there must members who are willing to devote their time to the administration. At the very minimum, associations need a chairperson, a secretary, a treasurer and a membership secretary. Depending on the breadth of the association's activities the association may also need a web master, an apiary manager, an education officer etc. It is absolutely necessary for people to step forward to fill these offices if associations are to be able to fulfil their important role in beekeeping. You don't have to be the most experienced beekeeper but the people who keep the associations running are just as important as those who pass on knowledge and try to instil beekeeping standards.

As with any association of people, beekeeping associations will often become a support system for each other. Members may become ill or old and need help to continue managing their bees. Beekeepers, in their first few years of beekeeping, often need someone to act as a mentor. The associations and the people in them simply become a circle of friends.

The BBKA is an umbrella organisation to which most of the county organisations in England are affiliated. It has various roles but the most important ones are to represent beekeeping to the media and government, to provide a liability insurance for all registered members, to provide training and to organise an annual convention.

BBKA includes within its structure an examination board which is largely independent of the rest of the BBKA. The examination board sets the standards for beekeeping throughout the UK by defining syllabi for beekeeping assessments from basic to advanced, both practical and theory and organises the assessments. There can be no doubt that by setting standards centrally it has been effective in ensuring that beekeepers in the UK are are as knowledgeable and skilled as any in the world. It could be accused of being rather dogmatic and inflexible in its approach to beekeeping and that its methods of assessment are not moving forward with the times, but this in no way devalues the achievements of the board, which must work within the constraints of monetary budgets and the skills and the energy of those that volunteer.

The functioning of the BBKA, like the local associations, depends upon individuals volunteering to take on the unpaid roles on the executive and on the examination board. The examination system also depends on a couple of hundred unpaid assessors.

As with all organisations that are managed by unpaid volunteers, on occasions it is not always run as effectively or as professionally as some might prefer. But those who criticise it have always got the option of standing for election to the executive committee or for the examination board.

The National Bee Unit (NBU), which is a part of the UK government department DEFRA runs a voluntary registration system, called BeeBase. Its primary purpose is to act as a database of beekeepers for the network of bee inspectors employed by the NBU that monitor bee diseases across the UK. By statute there are four diseases of honey bees which are designated as being notifiable diseases. The most significant of these are European Foul Brood (EFB) and American Foul Brood (AFB). Both of these diseases are highly infectious, and if left untreated, will kill a colony of honey bees and are highly likely to spread to neighbouring colonies. Any beekeeper that suspects the presence of a notifiable disease in one of their colonies is obliged to inform the NBU. Bee inspectors will then visit the apiary and take control of the measures necessary to confirm and then eliminate the disease outbreak. The action taken can include culling the infected colony. With AFB that is the only course of action that is feasible. Because of the highly infectious nature of these diseases the inspectors will also visit and check apiaries within a short radius of the centre of the outbreak and inform other beekeepers who are somewhat more removed, so that they can themselves check their bees.

The system works well, and outbreaks are usually confined and brought under control quite quickly. In the UK we are fortunate to have such an organisation in place. As a result, outbreaks of these devastating diseases are relatively rare and many beekeepers will never come across them in their beekeeping lives. The alternative would be to allow these diseases to become endemic and rely upon ongoing prophylactic treatment of the bees with antibiotics, as happens in some parts of the world. As a part of the measures to control disease the government have put in place restrictions and controls on the import of honey and honey bees. It is obvious however that the system relies upon each individual beekeeper cooperating and being fully engaged with these rules. Each beekeeper needs to regularly check their bees for notifiable diseases and notify the NBU unit when they suspect there may be a disease outbreak. During the spring and summer, two inspections each year should have the primary purpose of checking for brood diseases. For this to be effective beekeepers need to ensure that they are familiar with the symptoms of the notifiable diseases. And of course all beekeepers need to be registered on the BeeBase database. This is not mandatory and is a stark example of where ethics goes beyond what is required in law.

Every time a queen mates, a beekeeper could be relying on one of their neighbouring beekeepers or on feral bees for half of the genotype for their next generation of bees, and vice versa of course. So to keep bees that are bad tempered, or prone to disease, a beekeeper is not just compromising his or her beekeeping but also the beekeeping of neighbours.

As John Donne almost said, no beekeeper is an island. Well he might have said it if he had been a beekeeper. The point is that decisions that one beekeeper make effect his or her neighbouring beekeepers.

9 – Our Neighbours and Fellow Citizens

In this chapter there is a discussion as to how beekeeping activities may impinge upon non beekeeping neighbours and how the activities and mores of the wider society can effect honey bees and the pollinating insects.

When you keep bees you, at least to some extent, define yourself in your community. Your neighbours, even if they cannot remember your name, will know you as the person who has hives in the garden, sells honey or will collect swarms. Your beekeeping activities do not just impinge on your family, friends and other beekeepers, but also on the wider community.

Selling honey brings with it responsibilities. The authorities in England specify the quality standards for the honey that you sell in the Honey Regulations, which can be found on the internet. The honey regulations lay out these minimum standards. The most important of these is the one that relates to water content. Ordinary blossom honey must have a water content of less than 20%. There are a small number of exceptions to this rule, the most significant being that honey that is derived from ling heather may have a water content between 20% and 23%. Any beekeeper wishing to sell honey needs to measure and note the water content of each batch produced using a refractometer.

When the water content exceeds the stated limits it can very quickly begin to ferment. The surface of the honey in the jar will become dough like, there may be strings of carbon dioxide bubbles rising through it and it will have a characteristic boozy smell.

But the beekeeper wishing to sell honey needs to go beyond the honey regulations. The honey needs to be clear of impurities. If it is set or soft set the texture should be fine grained. If it is liquid it should be clear with no signs of incipient crystallisation. It should not have separated and of course it should taste good. The honey should be sold in good quality clean jars with clean lids and labelled in accordance with the regulations.

We live in a generation where the majority of people prefer to separate themselves from the natural world. They live in sterile homes, drive on metalled roads, travel in metal and plastic boxes over land, air and sea and work in large offices and factories where the air is conditioned. Beekeeping is a gateway to the natural world from which we have become divorced but with which many of us still yearn to be connected.

Because of the separation of modern lives from nature, there is a lack of understanding and a fear of stings in the general population which is not entirely rational. Being stung by a honey bee is quite rare as bees usually only sting in defence of their hive. People who never go near hives are very unlikely to be stung and the most likely people to be stung are beekeepers. It is true that stings are painful, but usually the pain lasts no more than a minute. It is also true that in a very small number of cases the sting of a honey bee can cause anaphylactic shock, the symptoms which include a difficulty in breathing and even a loss of consciousness. In extreme cases it can cause death. But the risk is small and should be put into proportion. In the UK, there are on average 34 deaths per year caused by skiing accidents, about 2000 people die each year in traffic accidents while just five people died last year from being stung by wasps, hornets or bees, only a very small proportion of which, if any, would be connected to beekeeping. In 25 years of beekeeping I have only seen one incident of anaphylactic shock which required medical intervention. The chap involved recovered quickly.

Having said that, if bees are kept in an urban or suburban environment there are precautions that must be taken to ensure that neighbours are not stung. Hives that are kept in gardens should be surrounded by a fence at least two metres high so that the bees leaving the hive are forced to gain height as they leave the hive, so minimising the risk of them coming into contact with people moving into the flight line of the bees. Where possible, hives should be sited so that they are at least 20 metres from a pedestrian right of way. The time when bees are most likely to interfere with neighbours is when the hives are opened for inspection. And therefore beekeepers need to keep their neighbours on side and let them know when the hives are being opened. Bees that are overly defensive should be replaced with a more placid strain, which can be done by replacing the queen. The gift of the odd pound of honey to neighbours is not misplaced.

Children are naturally fascinated by bees. It is the parents or teach-

Swarms don't always go where you expect.

ers that are more likely to express fear or squeamishness. Often beekeepers become messianic in their wish to promote beekeeping. It is only natural that beekeepers would wish to put together their desire to promote beekeeping with children's curiosity, interest and an urge to learn. It is perfectly possible to demonstrate the internal workings of a hive to children without significant risk of their being stung. Beekeepers often visit schools with observation hives and they are often in demand to speak to groups such as the women's institute and at craft fairs.

Bees swarm. No matter how conscientious you are in the way you manage your bees, you will lose the occasional swarm into your neighbourhood, but beekeepers should take all reasonable steps to prevent and control swarming. In a densely populated area, swarms may cause distress to neighbours, and so the beekeeper has a responsibility to minimise swarming and if swarming has occurred, the beekeeper must be prepared to collect it, provided this does not put the beekeeper or anyone else at risk. Swarms that are not collected may become a permanent nuisance.

When a swarm first leaves the parent colony it forms a cluster, usually a short distance from the parent colony. When the swarm is being cooperative it will cluster on a low hanging branch, and in this type of position it is easily collected by a beekeeper. Unfortunately, on some occasions, the cluster is less conveniently situated, for example on a fence post, deep in a thorn bush or high on a wall. While the swarm is in the cluster, scout bees are sent out to look for a permanent cavity which could be a hollow tree or in a chimney or the eaves of a house. To start with, the scouts may consider several possibilities, but eventually a consensus is reached. Typically this debate amongst the scouts may take two to three days. Then the swarm takes to the air and flies to its new home, guided by the scouts. It is generally true to say that people do not welcome colonies of bees into their chimney or in their eaves, since once they are there, there is usually no easy way to dislodge them without killing them. Though people do not want a colony of honey bees established in their house, normally they are high up and therefore not a nuisance to the humans who live there and they do not damage the fabric of the building. In many cases the human family are not even aware that they are sharing their home with a colony of bees. Nevertheless it is better all round if a swarm is collected when it is hanging in a cluster when it is still close to the hive from which it emerged.

Pollinating insects are an important part of the ecosystem, and honey bees are recognised as being one of the most important of these. The activities of these insects is something that benefits us all. We all eat the food from the crops that require insect pollination and we all enjoy the flora in the countryside and gardens that require pollination. Beekeepers are obviously contributing to this aim of preserving the population of pollinators but the responsibility for ensuring there are insect pollinators does not just belong to beekeepers but must be accepted by everyone. Farmers in particular, as guardians of large tracts of the countryside, have a very key role in ensuring that the population of the pollinating insects is not compromised. One of the most disturbing statistics that has emerged recently is the drastic 75% decrease in the biomass of flying insects in Germany over the last quarter of a century. The situation in the UK is likely to be similar. Many of us who belong to an older generation remember the way that car headlights would pick out thousands of insects when driving at night during the spring and summer, something that just doesn't happen these days.

The reasons for this decline in insect life are likely to be complicated and due to a cumulative effect of changes over many decades. One significant factor must be the way that our farmland is used and managed. When I was a boy I vividly remember lying in stubble fields at harvest time. The cereal crop had been under sown with clover and grass and in the flora that covered the field there was a teaming world of insect life of all shapes, sizes and colours. In those days no pesticides were used and disease and weed control was achieved through rotation. Now, arable farm land is continuously cropped year after year. There may be a change of crops but there is no rest for the soil. Pest control is achieved using pesticides, weed control by applying herbicides and fertility is maintained using fertilisers. Hedgerows have been grubbed up to make larger fields for the bigger tractors and agricultural machinery, marshy areas are drained and upland grasslands are converted into vast prairies. If you now lie on a stubble field there is bare soil between the cereal stalks, not an insect in view. Of course there has been a dramatic increase in the yields from UK farming and our food is produced much more cheaply. As a result the increasing human population of the UK eats well, too well in many cases, so that obesity and type 2 diabetes are becoming a norm. Food is so much cheaper. In the decade when I was brought up 33% of household expenditure was spent on food, now on average it is less than 15%.

It is wildlife in general, the flora and fauna, mammals, birds and insects that pay the price for the insatiable hunger of the human race. No politician dares say what should be said. If we want to conserve wildlife in all its forms the human race must consume less and breed less. But who will agree to either of these options. The developed world is hell bent to consume more, while the underdeveloped world continues to breed more. If we truly want our relationship with nature to be one of partnership and steward-ship then the human race will have to change direction. This must be the biggest ethical issue of our age, but one which is not yet being addressed in the political mainstream.

Depressingly the full solution is beyond any individual and even beyond any single government to implement. Farmers, as stewards of the countryside, have a vital role to play. Most farmers I know take this role conscientiously but they are primarily businessmen and their farm business must make a profit in a competitive envi-ronment. The use of pesticides is the most controversial issue. Insecticides are designed to kill insects and so if used carelessly they will kill our honey bees. There is a code of practice for spray-ing and it is important for the countryside that farmers and spray contractors comply with it. Every farmer I've ever spoken to takes it very seriously. Besides, sprays are expensive, and so it is in the farmers' interest to use sprays economically and avoid waste. In particular they need to avoid spray drift and endeavour to spray crops that are in blossom early in the morning before the insects are foraging. And if they are aware that there are beekeepers in the area they should make every effort to warn them in advance when spraying is going to occur.

In my opinion it is futile and unfair for beekeepers to indulge in a rant about greedy and uncaring farmers raping the countryside and killing the bees. Farmers in the main want to protect their land and they take pleasure in and value the wildlife they see about them. It is so easy to forget that we are all complicit in the drive to get more from each acre of land. We are complicit in populating our island with more people than the land can feed naturally and we like our cheap food. I wrote a few paragraphs before about remembering the rich insect life at the base of a stubble field. But I also remember how poor were the yields that were obtained then, compared to these modern times. As beekeepers we need to engage with farmers and make sure that when we are warned that spraying will occur then we shut the bees in the hives the night before so they are not effected.

There are many small things that can be done to preserve insect life and wildlife in general. Any of us with gardens can choose to have some wild areas rather than the pristine lawns, we can choose not to use chemicals , we can ask the local authorities that maintain parks to incorporate wild flower meadows into their designs, we can pressure local authorities to preserve wild areas along the verges of roads and motorways.

Some organisations that are involved with conservation of the countryside in the UK take the view that honey bees contribute towards the decline in the numbers of pollinators in the natural world. They suggest that honey bees should be regarded as domesticated animals that are artificially bred and when introduced into areas of wildlife conservation unfairly compete for nectar with other native pollinators such as bumblebees and so are complicit in the decline of native pollinators. Honey bees are also accused of carrying diseases that can spread into the populations of indigenous pollinators.

Beekeepers find this attitude difficult to understand and not helpful. It is true that, over the last fifty years, there has been a dramatic decrease in the numbers of indigenous pollinators and some species of pollinators are on the point of extinction. However honey bees are also indigenous to the UK. Fifty years ago, in the UK, there were significantly more honey bees colonies, but in addition there were a far greater number of other pollinating insects. Also beekeepers dispute the extent of the direct competition between honey bees and other indigenous pollinators as honey bees tend to be active much earlier in the year than most other pollinators and different species of bees have different lengths of proboscis, so are adapted to exploit different groups of flowers.

Beekeepers also maintain that the problems facing honey bees are shared with those facing other pollinators, such as the loss of suitable habitat and the use of pesticides. However it is clear that in countries where honey bees are not indigenous there have been damaging consequences to the native wildlife when they were introduced.

10 – The Economics of Beekeeping

Beekeeping can become a way to make a living or supplement one's income. In this chapter there is a discussion as to whether this is possible without effecting the welfare of the bees.

Most pastimes and hobbies involve expense and in this respect beekeeping is no exception. But beekeeping also presents the opportunity of making some income to offset the expenses incurred. For the small time hobby beekeeper there will usually be other motives to keep bees rather than profit. But when a beekeeper expands the number of colonies he is maintaining beyond three or four it becomes increasingly important that the income obtained is greater than the costs incurred.

The income will primarily come from the sale of honey. After the costs of the jars, the labels, the transport, a beekeeper does well to obtain a net profit of more than £3.50 per pound of honey. This will vary from region to region within the UK. The other factor that needs to be taken into account is the amount of honey that a hive is likely to produce. This varies greatly, from place to place, from year to year and colony to colony. I have had colonies that have yielded well over a 100lb of honey in a year, but I have also had colonies have produced nothing. In the Vale of York, over a period of years, I averaged about 40lb of honey per year per hive. This then would give a net income of £140 per hive. Of course this takes no account of the labour expended to manage the bees and

A honey stall at a farmer's market.

process this amount of honey. For a bee farmer, working alone, to achieve a turnover that is likely to result in an income equivalent to a living wage would require him or her to be looking after about 200 hives.

The costs of managing a single hive of bees are not insignificant. There is the depreciation on the value of the hive and equipment, sugar for Autumn feeding, medication, the cost of foundation, transport and honey processing costs. In total these can exceed £60 annually for each hive.

These figures may suggest there is a profit to be made from beekeeping. However statistics published by BBKA suggest that the average member of BBKA is achieving less than half the amount of honey that I quoted above and therefore many beekeepers are failing to make any positive margin on their beekeeping activities. For beekeepers who have just a few colonies this may be perfectly acceptable. After all it is a hobby and people are prepared to pay a cost for a hobby. But if each colony makes a financial loss, it is obviously a disincentive to expanding the number of colonies. It can be argued that, for the health and sustainability of beekeeping, it is necessary that in every area there are several beekeepers who are keeping larger numbers of colonies.

There are other ways in which beekeepers can make some return from their beekeeping. It is possible to sell other products of the hive, such as wax, propolis, pollen and royal jelly. Of these the most remunerative is the wax that can be made into candles and various ointments, soaps and cosmetics. But each hive normally produces less than a kilogram of surplus wax each year that is of good enough quality to be used in this way. Another source of income is selling bees and queens. It remains the case that by far the most significant income will come from honey.

Without there being some profit in keeping bees, beekeeping will become the preserve of the monied middle classes. Those who dismiss or at least minimise the importance of honey production from beekeeping may be excluding those who regard beekeeping as a harmless and interesting way of adding to their domestic income, even if in just a minor way. Our society depends upon us and our fellow citizens making a living, and there are very few amongst us that that truly claim that the income we are enjoying is not the result of exploiting other people or the natural resources in some way.

Once beekeeping becomes a money making exercise other issues come to the front. If beekeepers view making profit as having a higher priority than the welfare of the bees and their responsibilities to their customers and society in general that is undesirable. However there is no necessity that the desire to make beekeeping into a profitable pastime should impact on the wellbeing of the bees. Bees that are well cared for and in good health are more likely to produce good honey crops than those that are badly looked after.

A Final Note

Beekeepers come from all walks in life. I have known beekeepers who are farmers, teachers, firemen, railwaymen, hospital consultants, therapists, midwives, foundry men, clergymen, airline pilots, bankers, computer developers, office workers, businessmen, scientists. Beekeepers may be men or women, wealthy or poor, old or young. Beekeeping removes conventional hierarchies and establishes a new equality. I have seen hospital consultants go to a fireman for beekeeping advice.

There must be something that draws people from different backgrounds together as beekeepers, but it is not so easy to express exactly what that is. Beekeeping is an opportunity to intensely experience what we are, human beings in the round, connecting with the natural world and the other inhabitants of this wonderful world in which we live. Beekeeping reminds us that we have hands that can make things and carry out the most delicate of tasks rather than just type letters at a keyboard. It reminds us that we have strength in our arms and back. It reminds us that we have eyes to see things that we must interpret in our own way. It reminds us that we have minds that can work out things for ourselves from what we see and experience rather than absorbing second hand ideas from the television, radio, books, teachers and career advancement courses. It reminds us that there are things in this world that can inspire us with wonder and tap dormant reservoirs of creativity that we remember we had as children. Beekeeping awakens in us the curiosity and wonder of the child. When we

Beekeeping awakens in us the curiosity and wonder of the child.

are beekeeping we can cast aside the petty rituals of life, social competition and fashion. We become just ourselves with the bees.

To all the above, I add ethics. Ethics is an essential element of what it is to be human, to be able to judge what is right and wrong, and so it is only proper that beekeeping, so much a model of our humanity, should include ethical considerations.

There is always the danger that by highlighting where we can go wrong in beekeeping it might forever discourage some embarking on it in the first place. But then the same could be said of choosing to live with a partner, having children or choosing a career, where again there are so many opportunities to get things wrong. Yet we persevere and continue to wrestle with the ethics of these situations. And so I hope it should be with beekeeping.

Finally, a few sentences about me. It may be assumed that someone who chooses to write about ethics is blessed with certainties about life, but, sadly, this is not the case. As a child and working adult I had mathematics which acted as a security blanket of certainty. But with the rest of life I have struggled to find that certainty, never being absolutely confident as to what is right or wrong or where lines should be drawn. I see too many issues as being grey rather than black and white.

I started beekeeping about twenty five years ago.

At present I keep my colonies in national hives. In the past I also had a small number of colonies in Warré hives. I love my beekeeping. I have spent hours studying honeybees, thinking about them, knowing that as long as I wish it so there will be more to learn. I have an on-going sense of wonder that the simple but profound principles that Charles Darwin first set out a hundred and fifty years ago, should have resulted in a creature so complex in its behaviour and physiology, and so well adapted to its niche in the environment.

John Whitaker
Herefordshire
May 2018

johnmartinwhitaker@hotmail.com

Lightning Source UK Ltd.
Milton Keynes UK
UKHW05f1837220718
326102UK00002B/6/P